UNIVERSES WITHOUT US

UNIVERSES WITHOUT US

Posthuman Cosmologies in American Literature

MATTHEW A. TAYLOR

UNIVERSITY OF MINNESOTA PRESS
Minneapolis
London

Portions of the Introduction and chapter 1 appeared in "The Nature of Fear: Edgar Allan Poe's Posthuman Ecology," *American Literature* 84, no. 2 (2012): 353–80; copyright 2012 by Duke University Press, all rights reserved; reprinted by permission. Portions of chapter 1 appeared in "Edgar Allan Poe's (Meta)physics: A Pre-history of the Post-human," *Nineteenth-Century Literature* 62, no. 2 (2007): 193–221. Portions of chapter 2 appeared in "The Phantasmodesty of Henry Adams," *Common Knowledge* 15, no. 3 (2009): 373–94; copyright 2009 by Duke University Press, all rights reserved; reprinted by permission.

Published by the University of Minnesota Press
111 Third Avenue South, Suite 290
Minneapolis, MN 55401-2520
http://www.upress.umn.edu

Library of Congress Cataloging-in-Publication Data

Taylor, Matthew A., 1978–
Universes without us : posthuman cosmologies in American literature / Matthew A. Taylor.
Includes bibliographical references and index.
ISBN 978-0-8166-8060-3 (hc)—ISBN 978-0-8166-8061-0 (pb)
1. Cosmology in literature. 2. American literature—19th century—History and criticism. 3. American literature—20th century—History and criticism. 4. Poe, Edgar Allan, 1809–1849—Criticism and interpretation. 5. Adams, Henry, 1838–1918—Criticism and interpretation. 6. Chesnutt, Charles W. (Charles Waddell), 1858–1932.—Criticism and interpretation. 7. Hurston, Zora Neale—Criticism and interpretation. 8. Humanity in literature. 9. Human beings in literature. 10. Self in literature. 11. Order (Philosophy) in literature. I. Title. II. Title: Posthuman cosmologies in American literature.
PS217.P45T39 2013
810.9'384—dc23

2013028372

Printed in the United States of America on acid-free paper

20 19 18 17 16 15 14 13 10 9 8 7 6 5 4 3 2 1

CONTENTS

ACKNOWLEDGMENTS

My universe is populated with many wonderful people, without whom this project never would have been possible. At the project's inception, Sharon Cameron and Amanda Anderson offered unfailingly wise guidance, inspiration, and support, and for their willingness to read painfully early drafts I owe Michael Moon, Jennifer Culbert, Simon During, John Irwin, Hollis Robbins, and Ron Walters my deepest gratitude. My friends at Johns Hopkins University ensured that my arguments never received less than rigorous hearings; thanks in particular to Tara Bynum, Bryan Conn, Jason Hoppe, Claire Jarvis, James Kuzner, Mark Noble, and Dan Stout.

I could not be more fortunate than to work in the Department of English and Comparative Literature at the University of North Carolina–Chapel Hill, where colleagues and students alike are routinely amazing. For their collegiality and intellectual generosity, I especially thank Neel Ahuja, Dan Anderson, David Baker, James Coleman, Danielle Coriale, Tyler Curtain, Jane Danielewicz, Gregory Flaxman, Laura Halperin, Jennifer Ho, Heidi Kim, Shayne Legassie, Megan Matchinske, John McGowan, Eliza Richards, Jamie Rosenthal, Rebecka Rutledge Fisher, Beverly Taylor, James Thompson, Jane Thrailkill, and Henry Veggian. The students in my Posthumanism and Futures of Theory seminars proved to be game and insightful interlocutors at crucial moments in the development of the book's argument. I also thank Priscilla Wald (and all of the participants in the Duke Americanist Speaker Series), Cary Wolfe, Jennifer Fleissner, N. Katherine Hayles, Jack Kerkering, Caroline Levander, Barbara Herrnstein Smith, and Anne-Lise François for their challenges, encouragements, and kindnesses, large and small.

Much of the archival work for chapters 3–5 was conducted at the Library of Congress, which reference specialist Judith Gray ably helped me to navigate. This research was funded in part by a UNC-IBM Junior Faculty Development Award, and a departmental research assignment offered the time needed to bring the project to completion; in addition, a generous University Research Council Award assisted in the costs of securing image rights and preparing the manuscript. I am very grateful for all three. My graduate research assistant, Emma Calabrese, provided invaluable aid in tracking down last-minute sources and preparing the manuscript. At the University of Minnesota Press, the anonymous readers' incisive comments gave me a wonderful gift: the ability to see my argument anew. The book is much better for their interventions, which truly represented the ideal of peer review. Finally, Doug Armato's and Danielle Kasprzak's grace, good humor, and wise advice have made the entire process a genuine pleasure.

The support of my friends and family has meant everything throughout the years of this book's genesis; even when far away, they were there at every step. Carrie, Tim, and Barbara, you can't know how much your support has meant to me. To Elisha, most of all—encourager, wordsmith, and profoundest friend, thank you.

Introduction
IMMORTAL POSTMORTEMS

OF STONE, AN ALMOST HUMAN FORM. Legs and genitals mired in frozen earth. A single arm gripped in abbreviated expression, touching rock where hands and head and face should be. Almost a body, equally a mass, it is a thing imperfectly articulated from its medium, incapable of pulling itself out of indistinction. If it is a body, it is one that cannot be distinguished from its effacement—an identity, yes, but not its own. It is a suffocating expiration of the space between figure and marble, person and environment, part and whole.

It is Michelangelo's *Blockhead Slave* (Figure 1), and it has long been viewed as an icon of Renaissance thought, a materialization of the Neoplatonic-cum-Gnostic attempt to emancipate spirit from embodiment.[1] In this light, the sculpture is an allegory of the supposed human condition, where all of us are slaves to our earthly corporeality, all of us paradoxes of living noumena and dead phenomena. *Blockhead* here becomes the formal antithesis and necessary complement to the sculptor's *David* (Figure 2), who, perfect of design and clear of purpose, imperiously towers over base and world.[2] Indeed, *David* can be seen as *Blockhead*'s evolutionary telos, the ideal redemption of a nightmarish Babel of order and chaos: in *David*'s body is the successful delineation, spatial and ontological, of *Blockhead*'s failed remove. Two sculptures, two cosmologies. *David*: individual and representative, pinnacle of creation, human bridge between mundane and divine. *Blockhead,* also known as the *Atlas Slave*: a different kind of intermediary, Atlas, one who separates rather than brings together—and if, in this case, the burden proves too heavy to bear, if Atlas shrugs, then the result is not so much ecstatic communion as fatal collapse.

FIGURE 1. Michelangelo's *Blockhead* or *Atlas Slave* (1532). Accademia, Florence, Italy. Courtesy Alinari/Art Resource, New York.

FIGURE 2. Michelangelo's *David* (1501–4). Accademia, Florence, Italy. Courtesy Alinari/Art Resource, New York.

There is, however, another way to read the relationship between *Blockhead* and *David,* one that forgoes the narrative of progress from non-*finito* to *finito,* parts to set apart, slave to king, morass to Man. In this reading, *David* is neither more real nor more ideal than *Blockhead,* as the performative contradiction of the former's identity—stone embodying spirit—is foregrounded. Here, too, *Blockhead*'s lack of definition is less a failure to be transcended than an inevitability to be borne. Revealing the boundary between person and world to be something other than absolute, the Atlas maps a different set of coordinates than *David* by disintegrating *David*'s illusion of ontological liberation. For *Blockhead,* there is no masterly isolation to be won from the environment, no self-constitutive cleavage from context. Instead, *Blockhead* insists, the medium of our selves is also that of our unmaking, the world not a lifeless weight with which we are encumbered but the very condition of being, the vital stuff out of which we are both constructed and deconstructed. According to *Blockhead,* then, we are all of us Adam—grounded, of the earth—where the ground itself is alive with the unexpected, immanent agency to reclaim our selves, to erode the all-too-human separation that was only ever another form of contact.

PRODUCED FIVE HUNDRED YEARS LATER, James Cameron's *Avatar* echoes *Blockhead*'s integration of self and world, if not the sculpture's affect regarding it. Set in a future—conspicuously like our present—where the exhaustion of resources at home leads militarized corporations to invade foreign lands, the film offers through its protagonist, Jake Sully, a utopian alternative of ecological harmony. Jake, paralyzed below the waist during his prior service as a marine, participates in an experimental procedure that temporarily downloads his mind into a genetically engineered avatar of the Na'vi, the indigenous humanoid inhabitants of the extraterrestrial moon Pandora. He soon assimilates into Na'vi society, decides to fight with them against their would-be human exterminators, and finally marshals his forces to victory by uniting the moon's entire biosphere in a Gaian defense against the technologically superior invaders. In reward, the global energy-force Eywa, who inspirits all Pandoran life, transposes Jake's soul into his avatar, thus severing his dependence on the laboratory chamber that previously joined the two. Twice born again, he is at the end of the film the undisputed leader of the Na'vi, having transcended his base humanity and attained, in the words of Cameron, the status of "something that is our higher . . . or . . . aspirational selves."[3]

Such a transcendence makes the movie a fitting avatar of what has come to be known as posthumanism. Although commonly associated with near-future forms of technological augmentation—cybernetics, genetic engineering, mind–computer interfacing, and so on—posthumanism is best understood as a broader, transhistorical attempt to integrate the human into larger networks of being (it is this emphasis on positive integration that differentiates posthumanism from antihumanism's negative or deconstructive critique, evident in early Foucault, Lyotard, and Derrida).[4] By this definition, posthumanism encompasses not only technoscientific transhumanism (e.g., the initial porting of Jake's mind into the artificially manufactured Na'vi body) but also older, romantic forms of incorporation into natural environments (e.g., the Na'vi ability to connect psychically with Pandora's sentient flora and fauna via the symbiotic neural interfaces in their ponytails).[5] In both instances, Jake is made more than himself by abandoning himself, by escaping the boundedness of his human incarnation.

This flight into the more-than-human is *Avatar*'s answer to the villainous Resources Development Administration (RDA), whose ravaging of Pandora is rooted, the film suggests, in a cosmology that deems humanity to be ontologically distinct from and superior to the nonhuman world.[6] Agent of progress, rationality, and order, the RDA regards Pandora as mere chaotic stuff, raw material to be refined unto obliteration. The RDA, that is, exemplifies all of the humanist inhumanities so frequently laid at the feet of the Renaissance and Enlightenment philosophical tradition—which makes Jake, as demonstrated by his overcoming of this tradition's mind–body, human–nonhuman, and self–environment dualisms, its posthumanist correction.

And yet there is in *Avatar*, as in much posthumanism more generally, an uncanny reminder or remainder of the very humanism it attempts to supplant.[7] Jake's transformation from human disability to posthuman ability, for instance, intensifies rather than departs from humanism's belief in progressive perfectibility: his physical and cognitive capacities heightened, his individual agency amplified, his life extended—all confirming N. Katherine Hayles's charge that transhumanism merely "expand[s] [the] prerogatives" of the "autonomous liberal subject . . . into the realm of the posthuman."[8] His assimilation into the Pandoran ecology results not, as Bruno Latour claims, in a "redefin[ition]" of "what it means to have a body, a mind, and a world" but in a repackaged, self-serving anthropocentrism that imagines enmeshment to equal enhancement.[9] Although it is true, therefore, that Jake's euphoric convergence with Pandora ostensibly rejects Cartesian dualism, it is equally self-centered, constructing the self out of the world rather than out of an opposition to it; the world may no longer be our other but only because it has become our selves. It is now a carnival mirror reflecting our exaggerated, universalized images. As Adorno feared of Marx's positive dialectic, *Avatar* converts the world into "a giant workhouse" for the production of the self—an instrumentality not far removed from that of the RDA.[10]

Such a criticism may seem excessive in relation to *Avatar,* but this is the crux of the posthumanist gambit in general, whether post-human-ist (a form of superhumanism) or post-humanist (a positive form of antihumanism):[11] the apparent overcoming of the subject–world opposition threatens merely to consolidate it, the rapprochement achieved only by erasing the world as a phenomenon independent of ourselves. There is the potential, in other words, for a slippage from speaking of the self's assimilation *into* the environment to speaking of the self's assimilation *of* the environment, the dream of dissolving deleterious distinctions becoming the nightmare of an all-colonizing subjectivity.

We might have reason to pause, therefore, when posthumanist theorist and visual artist Robert Pepperell implicitly responds to Donna Haraway's question of "why should our bodies end at the skin?" by answering that, in fact, "human bodies have no boundaries.... There is nothing external to a human, because the extent of a human cannot be fixed." So, too, might we hesitate when Daniel White, author of *Postmodern Ecology,* draws on Humberto Maturana and Francisco Varela's work to conclude that the cosmos "forms our own extended identity" or when Paul Shepard, anticipating White by decades, argues that seeing ourselves "as part of the landscape and the ecosystem" means that "the self [is] ennobled and extended rather than threatened." Finally, we might fear the consequences when environmental philosopher Neil Evernden claims that "the extension of the boundary of the self into the 'environment'... imbue[s] it with life" and then approvingly quotes Northrop Frye's contention that "the goal of art is to 'recapture . . . that original lost sense of identity with our surroundings, where there is nothing outside the mind of man.'"[12]

On the basis of this evidence, gleaned from what might broadly be called posthumanist ecology, we apparently stand to gain more than a little from the loss of our bounded selves, as deconstructive play becomes an ontology of personal edification (Haraway, for instance, promises "*pleasure* in the confusion of boundaries," including the "pleasure of connection . . . [with] other living creatures" no longer regarded as our others).[13] But that we should expect to benefit from "the collapse of [the] ontological boundaries" that separate us from the world betrays an incredible degree of optimism, even anthropocentrism, regarding our posthuman condition, all the more so given it is a condition supposedly based in dynamism, contingency, and uncertainty rather than in older ecological models of harmony, equilibrium, and balance.[14] More disturbing, however, is the assumption that enables this optimism: in the absence of boundaries, it would appear, the self is infinitely extendable into the world, free "to become . . . with no origin or end."[15] Of course, the result of this expansion is that no outside is left, that nothing but the self remains ("there is nothing external to a human"; "there is nothing outside the mind of man"). In these pushes to identity, it is not our interiority that disappears but the externality of the world, which becomes merely the shadow of our now universalized selves. Thus, for many posthumanisms, the fact that "it becomes harder and harder to say where the world stops and the person begins" does not mean that persons go away.[16] On the contrary, "persons" now *become* the "world." "We"—or something uncannily similar—survive to enjoy the loss of our selves.

We might then say that the humanist self is formed by a constitutive exclusion of the world, whereas the posthumanist self is born of an ostensible inclusion of it, but the distinction is largely semantic: in both, the nonhuman world serves our all-too-human agendas, either by being our slave or by becoming ourselves. Put another way, posthumanism can uncannily echo the means–end logic of the anthropocentric metaphysics it supposedly corrects, ending the dialectic with otherness only to achieve a final synthesis of the self. Read in this light, the question "why should our bodies end at the skin?" sounds not ethically inclusive but disturbingly imperialist, as the world becomes only something to be colonized by our limitlessly possessive selves.

This is not to say that some posthumanist critics are unaware of such dangers. Haraway, Hayles, and Deleuze and Guattari, in particular, are quite cognizant of the ways in which the discourses of potentiality, emergence, and interconnection can be coopted by networks of "domination," and they all underscore the consequent need to "tak[e] responsibility" for the circuits and assemblages of which one is a part.[17] The combination of a negative critique with the positive articulation of posthuman being is a precarious one, however, and to the extent that it is even marginally defined by prescription rather than description, by what Andy Miah labels a "visionary stance," it threatens to arrogate for itself the ontological privilege of the humanism it seeks to overcome.[18] We would do well, therefore, to attend to the potential inhumanity of our becomings, to fear the possibility of "the *ecologization* of the subject" becoming a "*rhizomatics of domination*," even when such domination looks for all the world like love.[19]

It is a caution born of a history of affirmative self–world fusions at least as old as their supposed severance. In this sense, Latour is right: we have never been modern because we have never lived what the modern constitution mandates—the absolute divide between human subjects and nonhuman objects.[20] Even Bacon, Newton, and other luminaries of the new sciences exhibited a posthumanist strain insofar as they believed in humanity's subjection to occult powers of interconnection (astrology, witchcraft, the mysterious action-at-a-distance of gravity), secretly searching for the law that would unify all existence, spiritual and corporeal, physical and metaphysical.[21]

Of course, they searched for this law so that they might bring it under human control (Bacon contends that the end of knowledge is "the Enlarging of the bounds of Humane Empire, [for] the Effecting of all Things possible"), an imperial confidence built on the edifice of reason and born of the assumption that the universe culminated in Man.[22] By the uniquely human capacity for empirical science, it was claimed, the supernatural would be made natural; the paranormal, normal; the invisible, visible. Waiting to fulfill its destiny and relinquish its secrets, Nature would at last kneel, conquered and redeemed, before its rightful master.

No doubt this domineering attitude is part of the temperamental contrast between Enlightenment and romantic cosmologies, along with a more general epistemic shift from calculations of static mechanism to intuitions of dynamic organicism. Much romanticism certainly understood itself in this way, as correcting the previous era's deleterious instrumentalism and hyperrationalism. As Marshall Brown and others have argued, however, there are at least as many continuities as differences between the two.[23] Schelling's idealist *Naturphilosophie*, for instance, shared its predecessors' desire to unite the physical and the metaphysical, and romantic thought still largely extolled the human as the pinnacle of phenomenal existence, that which brought reflection, voice, and meaning to the natural world.[24] At least in this respect, the two systems are contrapuntal variations on a common anthropocentric theme.

Take the British romantics, long heralded as exponents of a counter-Enlightenment cosmology.[25] Explicitly opposing the estrangement of human subjects from their nonhuman environments, poets such as Wordsworth and Coleridge ameliorate the self–world divide via a universal pantheism: "the one life within us and abroad / Which meets all motion and becomes its soul," in Coleridge's formulation, the "intertwin[ing]" of "human soul[s]" with the "Wisdom and Spirit of the universe," in Wordsworth's.[26] Yet, as critics since Keats and Hazlitt have noted, these claims to cosmic kinship are premised on an idealist subsumption of world to ego, a metaphysical reduction that parallels not only the physical exploitation of the earth but also the subjugation of "naturally" passive women and racial others.[27] Far from refuting Enlightenment anthropocentrism and its exploitative relationship to the natural world, the romantic ideal of "knowing ourselves" as "parts and proportions of one wondrous whole" here works toward "the sublim[ity]"—and thus the continued exaltation—"of man."[28]

The same charge can be made against many American transcendentalists. Consider Emerson's "The Over-Soul" ("that Unity, that Over-soul, within which every man's particular being is contained and made one with all other"), Whitman's "Song of Myself" ("I know I am solid and sound / To me the converging objects of the universe perpetually flow / . . . I know I am deathless"), or Thoreau's "Walking" ("I believe that there is a subtile magnetism in Nature, which, if we unconsciously yield to it, will direct us aright").[29] All would seem to model a form of holistic interconnectedness that rebukes "degenerate" modernity's "discord . . . between man and nature."[30] But, as with their British romantic counterparts, such détentes are established only by subordinating the world to an occasion of the self's realization. Thus can Emerson claim in an uncollected poem that he does not "fear to die," for in death will be realized the full, self-exalting fusion of self and world only half consummated in life.[31] "The sanity of man," he confesses in "The Method of Nature," "needs the poise of this immanent force."[32]

On both sides of the Atlantic, the legacy of this contradiction would extend into the twentieth century, as multiple schemes of cosmic oneness altered the terms but not the fact of humanity's preeminence. Mesmerism, Swedenborgianism, orgone energy, Hegelianism, psychometry, spiritualism, popularized Native American and African religions, Hermeticism, Theosophy, and various New Thought movements all nominally disturbed the integrity of the individual human self by insisting on various forms of universal unity, but all simultaneously maintained that harnessing it would inevitably lead to the perfection of individuals, societies, or the whole of humanity, a fact evidenced by the myriad utopian communes and reform agendas associated with them. Our full potential, our true selves, our new heaven and Earth—each awaited only the bringing forth, the final realization, of the universal Spirit; our salvation wanted only our assumption of a godlike mastery already our own. As Rudolf Steiner claims in his 1918 *The Philosophy of Spiritual Activity,* "in reality I am the things [that I perceive] . . . insofar as I am part of the universal world process."[33] The self is not so much transcended, therefore, as it is infinitely expanded, its boundaries becoming those of the cosmos it replaces, the totality of the universe merely the intrinsic potential toward which the momentarily limited subject aspires. It is in this self-affirmative merging of subject and world that we can see a parallel, however unlikely, between optimistic counter-Enlightenments of the past and many present-day posthumanisms: in both, the effort to exceed the human self's limitations eventuates not in the self's abandonment but in its self-gratifying transformation; in both, we not only survive our integration into the larger world, we are, at last, *realized* by it.[34]

THERE IS, HOWEVER, another posthumanist tradition, one that reverses the division of human from cosmos without folding the latter into the former. In this alternative, whose queer form is intimated in *Blockhead,* both our separation from the universe and our identity with it are exposed as fantasies. Here we are incorporated into nonhuman processes, but the result is the erosion, not the augmentation, of our priority, with the attenuation of conventional human–nonhuman and subject–object boundaries culminating not in an all-consuming self but in a redistribution of agency and significance to the world.

It is this posthumanism that *Universes without Us: Posthuman Cosmologies in American Literature* explores through the works of Edgar Allan Poe, Henry Adams, Charles Chesnutt, and Zora Neale Hurston. Against the abiding nineteenth- and early-twentieth-century American faith in material and spiritual progress (whether through the externalization or internalization of "nature"), these authors picture universes that dethrone the human self from the apex of cosmic meaning.[35] Like their optimistic equivalents (of which they form an *immanent* development, remainder, or ghost), these cosmologies assume the existence of an all-encompassing law or principle, something that joins cosmic process and personal fate, but they contend that we are thereby endangered rather than ennobled; each, that is, presents a cosmology in which the cost of unity is our sovereignty. Refusing any self-affirming logic that would make the universe either serve or be us, these cosmologies thus disable the particular forms of mastery associated with Enlightenment and counter-Enlightenment selfhood: the mastery of otherness and the mastery of identity, respectively. If they reveal that owning the nonhuman universe is untenable because we are never above it, never outside of it, then they also show that being the universe is impossible because "we"—like *Blockhead*—are exceeded by the bonds of our becoming.

Read in this way, the cosmologies of Poe, Adams, Chesnutt, and Hurston are linked, via a principle of uncanny recurrence (i.e., the repeated return of humanism's repressed *post-*), to recent articulations of nonanthropocentric thought in Continental philosophy. Recognizing, for instance, power and vitality in nonhuman entities—both animate and inanimate, natural and artificial—normally denied such status, my authors preemptively answer Isabelle Stengers's call to avoid the "temptation to conceive of nature as submissive, manipulable, assimilable to some 'raw material' on which we would be free to impose whatever organization we choose."[36] Rather than inert nonbeing, the darkness against which our life is illumined, the world and the things that comprise it are in their cosmologies alive with the power to affect, resist, and, anticipating Quentin Meillassoux and other speculative realists, exist altogether independently of us.[37] In Jane Bennett's phrase, they picture cosmoses of "vibrant matter," where "vitality" is no longer an immaterial supplement possessed by a select few but the lowest common denominator of a newly flattened ontology.[38] On this side of dualism, a cosmological posthumanism.

Although enmeshment here restricts rather than extends our mastery, however, these cosmologies do not herald the victory of "nature" over "us" because they precede much recent ecocriticism in denaturing the false integrity of both the self and the idea of nature as essentially natural, other, or outside.[39] Inverting the Man–Nature binary simply reifies its terms, as does imagining a utopian conjunction of the two; to proclaim the rebirth of Nature upon the death of Man is to assume the anthropocentric fallacy of their dialectical distinction or synthesis. But the posthumanisms of Poe, Adams, Chesnutt, and Hurston abandon such static oppositions in favor of hybrid relations and dynamic human–nonhuman assemblages, where even the stability of things relating cedes, à la Whitehead, to the unfixed *event* of relation itself.[40] What results are spectrums of constructive and destructive interference rather than easy harmonies, with islands of order—body, person, world—revealed as temporary stays amid the oceans of chaos from which they arise and to which they return.[41]

It follows that in these cosmologies, as in much subsequent poststructuralist thought, the supposedly atomistic human self is run through with what was previously called otherness and still is something other than sameness: it is "a site of contamination and abduction by alien forces" (Keith Ansell Pearson), where "the subject is always already a *prosthetic* subject" (Cary Wolfe), but not only because of "the fundamental non-human character of language" (Paul de Man); in body, too, we see "the fundamental non-definition of the human as such."[42] In both senses—linguistic and material—the "I" within these universes is not alone but multiple and self-alienated, more transactional than essential, with externality at the very heart of its identity. Consequently, "self" becomes heuristically useful insofar as it demarcates an emergent, ephemeral property of nonhierarchical relations and deleterious to the extent that it assigns ontological purity to what is a much more contingent and co-prosthetic process. (And co-prosthetic because it is no longer clear where the subject ends and the object begins, the line having been reduced to one's temporal–spatial perspective; thus can we speak of a cosmos of quasi-objects and quasi-subjects.)[43] In both body and (embodied) mind, we are distributed, precariously open to our no longer external environments; what's more, according to Poe, Adams, Chesnutt, and Hurston, we always have been, which means that "we"—bounded, autonomous, self-determinative—have never been.[44] Far from existing in either imperious isolation or rapacious proximity, then, we see ourselves, like Poe's "Man That Was Used Up," as enmeshed with the nonhuman things within and around us—things both constitutive and deconstructive of "us."[45]

By putting Poe, Adams, Chesnutt, and Hurston into this unfamiliar context, *Universes without Us* significantly departs from previous scholarship, which has either ignored the cosmological dimensions of their works or read them allegorically—as mere vehicles for essentially human dramas of politics, race, gender, sex, economics, or personal psychology. Without denying their insights, my approach forgoes these hermeneutics of suspicion and symptom in favor of a more superficial or immanent method. Inspired by recent alternatives to critique that abandon prefabricated and often predictable explanations, I try to know with—rather than against or beyond—my authors' universes, taking seriously their consistent, if not systematic, depictions of the corrosive effects on human selves of radically nonhuman forces.[46]

In so doing, I follow Stanley Cavell, Sharon Cameron, Eve Sedgwick, Toril Moi, Cary Wolfe, Branka Arsić, and others in aiming to reactivate the philosophical potentiality of literature, a potentiality bracketed by the professional disciplining of knowledge over the century spanning Poe's and Hurston's careers (roughly 1835–1945). It was during this period that the academic separation of the humanities from the so-called natural sciences crystallized, a concrescence built on the parallel hardening of the distinction between human subjects and nonhuman objects.[47] As Lorraine Daston, Peter Galison, and others have argued, this threefold division—ontological, epistemological, institutional—became the foundational assumption of rational objectivity, of what would be knowledge and what *pseudo-*.[48] And yet, as evidenced by the countless theories of universal unity circulating at the time, it was a process whose positivist outcome became inevitable only in retrospect.[49] The final cleavage of scientific, religious, and literary concerns supposedly achieved during the period belies the messy intimacies of their actual interrelations.[50]

Unsurprisingly, then, one may note as a common characteristic of Poe's, Adams's, Chesnutt's, and Hurston's very different works the difficulty of extricating fact from fiction. Combining personal narrative and posthuman law, their generic hybrids (prose poem, scientific autobiography, plantation tales within realist frames, and autoethnography, respectively) attempt to capture truths unrepresentable by the rhetoric of objective detachment. Circumventing the contradiction of a view from nowhere that somehow escapes the universe it describes, these cosmologies narrate the effects of *inhabiting* cosmic process.[51] And because these are posthuman cosmoses, the effects include the deformation not only of the human subject but also of its disembodied, masterful vision.[52] Here cosmic forces affect narrative form as well as action, with personal and textual bodies alike succumbing before the pressures of the universe. As we shall see, there is a remarkable tendency for these texts to break down when they draw nearest their subjects, to collapse in exhaustion, madness, or death before apprehension can become comprehension. And rather than regard these collapses as failures of expression, we might consider them its consistent embodiment, indices of a certain brand of cosmic convergence. Seen in this light, the texts' very brokenness is communicative; where there is no detachment, where there is only the universe, the medium is the message.

From the perspective of current experimental sciences, these cosmologies would no doubt be dismissed as errant or "merely" fictional. My interest in them, however, is less in their empirical validity or successful depiction of reality than in the ways they anticipate, to a surprising extent, the concerns of an increasingly prominent strand of early-twenty-first-century philosophy. And yet I do not see the texts of Poe, Adams, Chesnutt, and Hurston as merely foreshadowing a current theoretical trend; I am not mapping the evolution of posthumanist thought from inchoate germ to present flowering. Rather, I treat these texts as fully realized in their own right, as agents—and not merely objects—of philosophical inquiry; neither prophetic illustrations nor palimpsests on which recent paradigms can be unproblematically applied, they are things to think *with*—even things with which to think the limits of thought—cointerlocutors in a historically and generically expanded posthumanism. In this way, *Universes without Us* introduces—sometimes as complement, sometimes as preemptive challenge—heterodox literary cosmologies into a posthumanist "ecology of practices" that includes ecocriticism, science studies, neovitalism, process-based materialism, religious studies, queer theory, animal studies, and critical race theory.[53] The result is less a genealogy than a series of untimely, mutually illuminative encounters that envision alternative pasts, presents, and futures of the self–world imaginary.

THUS FAR I HAVE PRESENTED my authors' relation to posthumanism in aggregate. In the chapters that follow, I differentiate among their respective visions by putting each into conversation with both an optimistic cosmology of its own era and a cognate strain of posthumanism from our own. Chapter 1, for instance, identifies a heretofore underrecognized cosmology manifest throughout Poe's short fiction and criticism—and explicitly theorized in the late proem-essay *Eureka* (1848). This cosmology shares with concurrent discourses of transcendentalism and mesmerism a conviction that the material and spiritual worlds are governed by a single law that transforms ostensibly discrete phenomena into elements of a universally interconnected whole—what "Ligeia" calls "a great will pervading all things."[54] Rather than elevating or empowering the human subject, however, cosmic unity for Poe signals an ominous fate, one that promises to erase our very existence. Indeed, according to *Eureka,* the entire universe is "in a state of *progressive collapse,*" the result of currently "disunited atoms . . . return[ing] into *One*"—as manifest in the law of gravity—so that they can reconstitute the God whose original act of self-diffusion created the cosmos.[55] From this inhuman perspective, *Eureka* can coolly regard the overwhelming of "individual identit[ies]" as "the most sublime of poems."[56]

What differentiates this vision from its numerous optimistic equivalents is that here "we" do not survive our divine convergence; "we" are an inevitable casualty of the drive to "imparticularity" because, for Poe, to be God is for one's self to be destroyed.[57] Neither can we draw any benefit in the present from our knowledge of the coming singularity. Portion of God though we may be, our portion in "the awful Present" is only the endurance of "Sorrow" and "Divine Injustice."[58] For the characters of the tales, therefore, who are subject to the same forces but not privy to *Eureka*'s knowledge of their telos (unless in a trance, like Valdemar, or morbidly hypersensitive, like Roderick Usher), there is only death, only the horror of their supposedly discrete selves being obliterated by the "terrible influence" of the universe collapsing around them.[59] The general permeability of persons to benignly generative forces in *Naturphilosophie,* metempsychosis, and Humboldtian ecology thus becomes, in Poe's redaction, a mortal danger manifest everywhere in matters more quotidian than sublime. Routing my discussion through mid-century treatises on mesmerism, the Deleuzian fold, the new materialisms, Heidegger's discourse on tools, Bill Brown's thing theory, and object-oriented ontology, I demonstrate that Poe's hylozoic things (from cats to houses to corpses to teeth) synecdochically embody his perverse version of the metaphysics of his age, with characters being destroyed by the cosmically invested, inrushing objects of their putatively everyday worlds.

Henry Adams only checked out one group of books during his freshman year at Harvard: Poe's collected works.[60] This may be nothing more than an odd bit of trivia, but my second chapter argues that Adams's texts evidence a similarly fraught dynamic between an author and the optimistic cosmologies of his age. I specifically contend that the famously uninhabited form of *The Education of Henry Adams* (1918)—its attenuation of author and subject via the third-person narration of "Henry Adams" as a "passive" and "helpless victim" of the "forces" surrounding him—registers Adams's attempt to reduce, in the mode of his contemporaries, both individuals and societies to the operation of deterministic universal laws.[61] Yet contra the anthropocentric evolutionary theories (e.g., Haeckel, Spencer, Fiske) and teleological historiographies (e.g., Hegel, Comte, Buckle) that made determinism a guarantee of humanity's future perfection, *The Education*'s "dynamic theory of history" and Adams's later "scientific histories" are considerably less sanguine regarding our prospects. Seeing no evidence of progress, either across races (as did Lewis Henry Morgan and E. B. Tylor) or uniformly as a species, but plenty of indications of coming catastrophe, Adams ultimately concludes that our existence is compelled by the annihilative principle of entropy, warning in *The Education* and elsewhere that because "chaos [is] the law of nature," we—and the very idea of "we"—face imminent "extinction."[62]

The Education's negation of its author thus captures in microcosm the self-obliterative energies animating, in Adams's view, the universe as a whole. Consequently, we are compelled to regard *The Education*'s oft-noted "irony" not as a studied aesthetic device, a psychological tic, or a component of a reformist politics but instead, with Kierkegaard and de Man, as the echo of the fictionality of one's self before the universe. Yet, for Adams, this recognition is neither theological nor poststructuralist—both potentially liberating—but terminally materialist, depriving us of a secure existential remove from which we might redeem ourselves.

In this he exceeds even his naturalist contemporaries. Although naturalism's universe is famously inhuman—"huge, furious, implacable" in Crane's depiction, "savage, ferocious, alive" in London's—it frequently is presented in the hope of producing a mitigating human response.[63] For Adams, on the contrary, universal force forecloses the possibility of any amelioration of our condition, and he underscores the point by repeatedly insisting on the utter nonproductivity of his theories. Similarly, whereas the formal structure of naturalist narratives testifies to the possibility of escaping the degradations they depict (with third-person narrators often removed from the scenes they survey), *The Education* spares neither its narrator nor itself from the anarchic energies of the cosmos, as both dissolve into nonhuman formulae. And lest we assume that by reading his life, we live to escape it, *The Education* repeatedly promises that Adams's fate is our own. The chapter concludes by exploring how this fatal consistency both resonates with and challenges recent philosophical inquiries into the place of the human (and the humanities) within nonhuman spatiotemporal registers, including Manuel DeLanda's inorganic evolutionism, Elizabeth Grosz's "active" time, and Wai Chee Dimock's deep time (as well as its radical extension in the extinction theories of Mark McGurl and Claire Colebrook).

As much as Adams believed his "neck [to be] broken" by the "anarchical" cosmic powers he surveyed, his patrician status ensured that, at least socially, he was insulated from the full force of their devastations.[64] Indeed, both of my first two chapters examine white, male authors who denatured models of universal subjectivity that were their natural inheritance, but who did so in ways that could be seen as evading or even furthering what Russ Castronovo calls the disembodied ideal of "necro citizenship."[65] In the very act of bringing the cosmos down around them, Poe and Adams arguably confirm, inversely and extravagantly, anthropocentric white male privilege. My next three chapters, however, turn to those excluded from such privilege, those who were considered things so that others might be men: literal quasi-objects–quasi-subjects. More specifically, these chapters explore the dehumanization of African Americans in chattel slavery and its racist afterlives—where persons were made prostheses in the fantasies of others' autonomy—but their real focus is on the African diasporic cosmologies that gave it the lie. This dynamic—the blindingly stark contrast between a fraudulent humanist universalism and a posthumanist cosmos of alternative equalities—is a primary reason that *Universes without Us* focuses on the U.S. context, even as it attends to the transatlantic and hemispheric circulations of bodies and ideas from which this context arose.

In the first of this triptych of chapters (an introduction to the book's second half), I consider influential accounts that read certain African American slave autobiographies as posing an "Africentric alternative" (Paul Gilroy) to "the metaphysical system" of the West (Henry Louis Gates Jr.). While agreeing that these autobiographers confounded the taxonomies that naturalized the slave as more thing than human being, abject but never a subject, I follow Srinivas Aravamudan, Valerie Smith, and Saidiya Hartman in challenging the notion that they did so by embracing an African cosmology rather than by extending an Enlightenment one to include themselves. Studying Olaudah Equiano's *Narrative* and Frederick Douglass's *My Bondage and My*

Freedom, I explore how these men demanded their citizenship—national, religious, human—by pointedly renouncing Africanist beliefs, thereby proving their status as rational, self-determinative individuals. This was neither a betrayal nor a mistake: the social and political imperatives of the time necessitated that they do so, that they gain recognition as persons to end their immediate subjugation, and the continued relevance of this precedent is obvious. The very attempt exploded any pretense to their inequality. But equality with one's enslavers, though necessary, comes at a cost, as demonstrated by Douglass's assertion that he is a man because he can dominate both other men and the nonhuman world.

There are, however, other ways to controvert the metaphysics of racism, and as a way of setting up the second and third of these interrelated chapters, I draw on Works Progress Administration archives and scholarship attesting to the New World survival of African diasporic cosmologies to identify two alternative sites for their manifestation: Chesnutt's *The Conjure Woman* and Hurston's *Mules and Men,* both of which picture conjure, hoodoo, or voodoo as overwhelming the distinctions not only between black and white but also between human and nonhuman, subject and object.[66] Such attenuations might seem to promise an enabling conception of the self's relation to the world. Indeed, this is how voodoo in particular was received during the latter portion of the U.S. occupation of Haiti (1915–34) by many northern white elites, who regarded it as an exotic counterpart to other contemporary "alternative" spiritualities that joyously announced the self's sublimation within a spiritually endowed universe.[67] When it wasn't castigated as devil worship or ignorant superstition (two contradictory attitudes equally on display in such popular fare as 1932's *White Zombie*), voodoo was folded into a cosmology that sought to abandon Enlightenment individualism even as it promised intensely individual pleasures. Despite frequently being antiracist, therefore, such accounts perpetuated the logic that the nonhuman naturally served the human. In these popularizations (including Pauline Hopkins's Ethiopianist *Of One Blood*) as well as in the increasingly prominent demand for mass-market conjure or voodoo items, cosmic energies were controllable potentialities waiting to fulfill us, and fusing oneself with the cosmos meant merely accessing its powers as one's own. The self was "lost" to the universe only to be miraculously reborn, new and improved.

But such representations do not exhaust conjure's or voodoo's potential to destabilize forms of domination based on absolute human–nonhuman and self–other distinctions. As in the African traditional religions informing them (and in the contemporary neoanimisms and actor–network theory that they, in turn, precede), there is in conjure and voodoo an extension of consciousness to "inanimate" places and things; an attenuated category of the human; and diminished or decentered senses of individual identity. And in Chesnutt's and Hurston's texts, in particular, this radical permeability to the universe leads to self-abnegation rather than self-sovereignty. Inevitably focalized through the experiences of New World slavery (but never wholly reducible to them), the self's openness to the cosmos frequently equates with its fracture: lizards beneath the skin; bodily metamorphosis; possession by spirits; death by voodoo dolls. Like gravity in Poe's universe and entropy in Adams's, universal energies are accessed only at the cost of "one's" life, as becoming a part of the inhuman universe means coming apart as a distinct, coherent person.

By this account, certain African diasporic spiritualities refute racist ontologies but do not thereby liberate us as free subjects, disallowing Ethiopianism by the same principle that they reject white cultural nationalism (the period between Chesnutt's and Hurston's major works saw the passage of most Jim Crow laws, the height of southern lynching, the "benevolent assimilation" of the Spanish–American War territories as well as Haiti, and the rise of the second Ku Klux Klan). This is because the decentering force of the voodoo–conjure cosmology exceeds its disruption of racial binaries to dissolve the integral, humanist self as well, making it a casualty of the abandoned self–world boundary. Indeed, the impossibility of owning one's self under slavery becomes catholic under voodoo, where all supposed masters are made slaves to their condition—with the crucial difference, of course, being that "dehumanization" in the former results from an artificial exclusion to be overcome, whereas in the latter it follows from a universal law that cannot be.

Far from romanticizing slavery, this cosmology explodes its ontological justifications by leveling slave, master, and nonhuman world. Consequently, although the cosmology utterly incapacitates the racist distinctions that would regard African Americans as somehow unequal to the standard of white humanity, it does so not by proving that African Americans meet this standard (as did many slave autobiographies) but by dismissing it altogether. Equality is still insisted on, but it is predicated on a new set of metaphysical coordinates that forgo the absolute distinction between human and nonhuman. Anything more or less would be, like biopolitical racism, a self-serving, other-excluding mystification, a violent instrumentalization of a nonhierarchical difference.

Chesnutt's tales, for instance, show conjure—much more tangible than Adams's entropy—enabling a radical integration with the natural world, as characters are transformed into aspects of their physical environments (trees, wolves, grapes, etc.); but rather than such proximity being a sublime experience or evidence of a self-serving "universal beneficence," as in John Burroughs's contemporary nature writings, it is inevitably annihilative of those who endure it.[68] Thus, drawing on the posthumanist ecocriticism of Timothy Morton and others, chapter 4 puts Chesnutt's tales into an interrogatory dialogue with the self-expansive protoenvironmentalisms of his era as well as with current scholarship that reads Chesnutt's fiction as affirming them. It also examines the tension between the folk tales' corrosion of anthropocentric individualism and the humanist frame narratives circumscribing them (as well as the tension between the latter and Chesnutt's later domestic fiction), where discrete, self-interested subjectivity is reintroduced for the imperatives of an actionable politics. Ultimately, the two are irreconcilable, as Uncle Julius's victories in the frame narratives over his white employer presume a version of agential selfhood that is enabled by the *telling* of the conjure tales but disallowed by their *content*.

As a regionalist record of the oral histories and cultural dynamics of the postbellum South, Chesnutt's conjure fiction is a precursor to the kind of anthropology later practiced by Hurston.[69] And like Chesnutt's, Hurston's depictions of hoodoo and voodoo focus on how the power of charms and spells to determine people's lives presumes the oscillation-unto-indistinction between subjects and objects. Yet, unlike the textually thin Aunt Peggy in *The Conjure Woman,* in *Mules and Men* (and the later *Tell My Horse*), it is Hurston herself who becomes indoctrinated into the secrets of hoodoo—to nearly lethal effect, her own life thinned to the point of disappearance. During one ritual in particular, the ethnographic enterprise breaks down as she is overwhelmed by forces incompatible with her coherent self, leaving her dazed, fearful, and silent.

This was not what Hurston's patron, Charlotte Mason, expected when she commissioned Hurston to study voodoo in New Orleans. A devotee of various occult beliefs, Mason saw voodoo as an ecstatically liberating practice, but Hurston's experience insists that voodoo is irreducible to such self-serving ends. In Hurston's account, one does not possess voodoo; one is possessed by it, with the result being the destabilization, not confirmation, of the autonomous self, even for the voodoo priests and priestesses who ostensibly wield its powers. Indeed, Hurston eventually dropped her voodoo studies for fear of her life. Years later, she would claim in her autobiography, *Dust Tracks on a Road,* that the "power" and "secrets" of voodoo ultimately would be explainable through "science" rather than "ceremony," but "it took months for [her] to doubt" her experiences in this way.[70] Achieving this distance was necessary, however, for Hurston to maintain the famously strong individuality undergirding her most lauded writings, including *Their Eyes Were Watching God.* Yet it also speaks to the fact that the hoodoo universe, unlike Chesnutt's rendition of it or Poe's and Adams's own cosmoses, demands only our sovereignty, not necessarily our actual lives. Indeed, Hurston's hoodoo suggests that our bonds to otherness, simultaneously constitutive and alienating, are the very flesh of life. Throughout the chapter, I compare this tension between an obliterative and a transformative in-corpo-ration of alterity with various phenomenologies of spiritual experience, as well as with Heidegger's concept of *Sorge,* Maurice Merleau-Ponty's "flesh," and David Abram's more recent account of the sensuousness of perception. Rather than subordinating Hurston's vision to this tradition, however, I examine how the hoodoo universe articulates a radically nonanthropocentric revision of it. In so doing, hoodoo conjures forth the perhaps impossible specter of a posthumanist ethics.

POE, ADAMS, CHESNUTT, AND HURSTON: an unlikely grouping, whether from the perspective of literary criticism or that of posthumanist theory. *Universes without Us,* however, endeavors to show how each author—in different historical contexts and with different philosophical consequences—pictures a cosmos that poses a threat to "us": Poe's cataclysmic corporealization of mesmerism's and transcendentalism's ecstatic unities, where persons are overcome by the hylozoic things accreting to them (prospectively engaging posthumanist univocity and thing theory); Adams's inversion of optimistic evolutionism, where entropic dissipation replaces Poe's perverse law of spiritual conservation (prospectively engaging posthumanist theories of deep time and extinction); Chesnutt's conjuring of early environmentalist rhetoric such that becoming one with the landscape ends in degradation, not colonization (prospectively engaging posthumanist ecocriticism); and Hurston's reflexive hoodoo ethnography, in which the mastery of both anthropology and popularized diasporic spiritualities is incapacitated along with their practitioners (prospectively engaging posthumanist phenomenology).

All these cosmologies destabilize, in their different ways, the ontologically privileged human self by imbricating it with nonhuman processes, but there is a further, crucial distinction to be made among them. If, in Poe's texts, things fall together (unmaking the self through addition), and in Adams's things fall apart (unmaking it through subtraction), then in Chesnutt's and Hurston's works we might say that things change (unmaking it through substitution or transformation), a fact that underscores the open-endedness of their universes against Poe's and Adams's closed teleologies. Whereas Poe and Adams see the future as already scripted, the deaths of humanity and universe both inevitable and coincident, the conjure–hoodoo–voodoo cosmology posits only the extinction of autonomous, discrete subjectivity, in the absence of which both cosmos and (reconfigured) persons continue to exist. The difference isn't one of science versus religion per se (both Poe and Adams regard the two as necessarily imbricated, and Chesnutt's conjure and Hurtson's voodoo include many naturalistic insights), but the latter posit an ongoingness that requires continued navigation for better or worse outcomes rather than the former's undeviating fatalism. The forces in the diasporic cosmos are still all-encompassing, still beyond our mastery, but now they are heterogeneous, even antagonistic, as myriad spirits, gods, charms, and countercharms interact in unceasing, dynamic interplay; interconnectedness remains, but it is here a principle of generative difference as well as destruction. It is the basis of life as well as death, and its relentless proliferations of difference render totalization impossible. The determinism of Chesnutt's and Hurston's texts, in other words, is undecided, circumstantial, and reversible rather than fixed, predetermined, and unidirectional—which is to say, it's hardly a determinism at all.

This is even more evident in Hurston's writings than in Chesnutt's, where the antebellum setting of Julius's tales means that the diasporic cosmology most often results in the death of actual persons as well as the death of metaphysical personhood. Hurston, too, reveals the violence of this cosmos, but in her ethnographies, what is endangered is less individual people than the assumption that people are truly individual, truly separate from or superior to the worlds within and around them. Lost with our possession by nonhuman agencies is the notion that we are, with Locke, self-possessed. Yet Hurston doesn't stop there; inspired by actual voodoo practices in New Orleans and Haiti, she explores the new opportunities and constraints that arise with the belief that existence is defined by relation rather than isolation, by mutual becomings of common flesh rather than discrete beings. Here anything—a plant, an animal, a place, a tool—could have a claim on "us"; indeed, "we" (and "they") might be understood as the sum of such claims, existing through our bonds rather than despite them: "we" defined as an action or a circulation, an emerging, ever-shifting constellation of reciprocal obligations rather than as a static and restricted set of impervious entities. The result is a simultaneously productive and terrifying uncertainty over the boundaries of our actions, our bodies, ourselves. Such an uncertainty can neither compel ethical action nor ensure harmony, but it can denaturalize the onto-logics by which racist and speciesist hierarchies justify the constitutive violence of their being.

What Hurston's example illustrates is that posthumanism need not be inherently pessimistic or negative any more than self-affirming. It would be merely another version of anthropocentrism to believe that the cosmos is innately misanthropic.[71] Thus the arc of my chapters: from Poe's and Adams's respective extinction scenarios, through Chesnutt's thanatopolitics of slave-era conjure, to Hurston's comparatively livable representation of hoodoo. Though each reproblematizes our settled answers regarding the line between self and not-self (reversing, in Latour's parlance, the premature conversion of this living matter of concern into a dead matter of fact), only Hurston offers the prospect of a future, posthumanist "us." The Latourian question I pose in the coda, extrapolating from Hurston's hoodoo, is whether it is possible to imagine any affirmative relation to this future without smuggling in a version of human exceptionalism or mastery that would violently coopt it. What, that is, might a nonannihilative, nonliteral posthumanism look like? Between salvation and apocalypse, one and zero, can we actually reconceive—in the light of my authors' cosmologies—what is meant by "we"? Can "we" be, or become, otherwise? Can there be a posthuman collective no longer bound to the violence of either total sameness or total difference? Can there be, to borrow from Christopher Castiglia, a politics beyond interiority or, to echo Spivak, a global-yet-plural ethics of "planetarity"?[72] Can a universe exist in which being, agency, and responsibility neither exclude nor are reducible to "us"?

It very well may be impossible, as even an ethics of restraint or care can presuppose the power, rather than the susceptibility, of the subject. This, after all, is Barbara Johnson's explicit critique of the "Levinas effect" (and thus her implicit critique of Judith Butler).[73] Yet this may be where my authors' cosmologies are most provocative: rather than finally reintroducing a form of human control over the nonhuman world, they all suggest that a fearful recognition of the human costs of living in an ahuman, indifferent universe may not be inappropriate; indeed, it could be a necessary check on the impulse toward self-universalization, and it could do so without reestablishing the humanist boundaries that separate us from the world. We can imagine a fear that would remain suspended in the apprehension of vulnerability, unable to vanquish its object because unable to regard it as either fully self or fully other—a fear that would recognize the self's constitutive integration with its environment without the ability to overcome it, thereby precluding reactionary, destructive attempts at mastery. Rather than reinscribing a defensive or sublime dualism between one's self and one's context, such a fear would be the inhabitation of a perpetually uncertain openness to the world; unable to fix "our" borders, we would find ourselves—if at all—incapable of establishing either a dominating remove or a self-serving identity. Neither utopian nor dystopian, possible nor dismissible, it would be an ontological quietism born of the terror of actions and identities never fully our own.

If, then, these cosmologies afford no ready-made politics or ethics, no easy answers as to how we are to live—indeed, if inhabiting these universes seems unimaginable—then that is part of the point, one source of their interest: these universes defy our attempts to remake them in our image; they offer—however fleetingly—a vision of a cosmos not our own, one that does not abide any naturalization of ethics or identity. It is a vision not long to be endured, perhaps even to be feared, but we know that we have seen it if in it "our" selves are less than assured, if, following Adorno, what we glimpse we can neither identify with nor reject, neither fully comprehend nor control.[74] Admittedly, this is a devil's bargain: rather than dominating the nonhuman world, we are now lost within it. But in attenuating our attachment to the absoluteness of human–nonhuman and self–other divides, these cosmoses at least dispense with a concept of Being that necessitates a violent defense of its boundaries. Offering neither a total war nor a lasting peace, they only whisper that something other than "us" might be.

1

EDGAR ALLAN POE'S META/PHYSICS

IN 1852, Herbert Mayo, renowned professor of physiology and anatomy at King's College, London, published in Philadelphia his *Popular Superstitions, and the Truths Contained therein, with an Account of Mesmerism,* which claimed to make all phenomena—from the "singular facts" of divining rods, clairvoyant trances, and vampires to the more common matters of batteries, magnets, and dreams—comprehensible through reference to a single, recently discovered energy allied to but distinct from electricity, magnetism, and heat.[1] This "Od" or "Odic force," as it was named by its discoverer, the noted geologist and chemist Baron von Reichenbach, supposedly animated the entire material and spiritual universe. Indeed, Mayo sought to convince his readers that the force's influence was ubiquitous, governing all things animate and inanimate, seen and unseen. But most important to Mayo was his more specific claim that individual persons were powerfully affected by the force; no less than the entirety of people's physical and spiritual lives, their strengths and weaknesses, successes and failures, fates and character traits, were made dependent on the particular amount and polarity of Od force around and within their bodies. Even consciousness itself, because predicated on electrical activity in the nervous system, was found to result from the Od force's action on the self. Putting oneself into harmony with Od energy, therefore, was akin to achieving a form of secular salvation; conversely, being in discord with it could explain any conceivable trouble.

Far from a cultural aberration, Mayo's study was representative of a significant current of nineteenth-century thought. Writing in the context of numerous theories premised on the existence of an omnipresent force or principle—including philosophical systems (panpsychism, Schelling's *Naturphilosophie,* Fichte's *Wissenschaftslehre,* Hegelianism, American transcendentalism), speculative sciences (vitalistic electrophysiology, Humboldtian cosmology), and cultural practices that conflated the two (mesmerism, spiritualism)—Mayo was committed to uncovering a rational, unified, and *material* model of the physical and metaphysical universe, as well as the place of the newly consolidated human within it. Of course, the Western idea that the whole of existence is reducible to a single law had a long history by this point, ranging from pre-Socratic atomism to Newtonian celestial mechanism. What the nineteenth century adds to this narrative is, on one hand, a romantic discourse of the individual's continuity with the "natural" world (and thus with the Spirit behind it) and, on the other, an increasingly professionalized scientific method enlisted to analyze these claims. In this context, we can see such mundane matters as the mid-century germ theory debates and Schwann's 1839 postulation of cell membrane permeability as participating in the same general conversation as Thoreau's *Walden* (1854) or the Fox sisters' 1848 spirit rappings: all evidence that bodies (and thus selves) are radically open to their surroundings, whether physical or spiritual.

For his part, Mayo regarded mesmerism as an established fact. He even considered it humanity's purest distillation of the Od force as of yet, certainly the most illustrative, because it perfectly captured people's dependence on "external" powers.[2] That this was so, that mesmerism's spectacular effects—telepathy, physical insensitivity to stimuli, reading through fingertips, etc.—were the result of putative individuals becoming subject to an apparently extrinsic determinant, was captured for Mayo and others in the entranced patient's ready acquiescence to the mesmerist's commands.[3] Moreover, because, for Mayo, the force behind mesmerism was not uniquely invested in its human participants but instead permeated all of existence, everyday places and objects were imparted with a new and extraordinary significance: they could upset or restore a person's Od balance, with baleful or beneficial results.[4]

For Mayo, then, mesmerism's revelation of the continuity between ostensibly differentiated persons and their environments was not merely incidental; it was the practice's condition of possibility. The metaphysical consequence of this deindividualization was that humanity was no longer a creation apart, an isolated point of significance within an essentially meaningless cosmos. And yet this integration did not, for Mayo, diminish humanity's standing. Conceding that the Od's operations were largely uncontrolled at present, Mayo had every confidence that humans would one day manipulate its energies for the improvement of the race. Indeed, Mayo intimated that mesmeric healings might be the first tentative step toward a future utopia in which a harnessed Od force would realize all of humanity's now dormant potentialities.

Like Mayo, Edgar Allan Poe actively studied mesmerism in the 1830s and 1840s, getting swept up in the American renaissance of a movement that had first come to prominence in Austria, France, and England in the 1780s.[5] In addition to writing three mesmeric tales ("A Tale of the Ragged Mountains" [1844], "Mesmeric Revelation" [1844], and "The Facts in the Case of M. Valdemar" [1845]), Poe reviewed and published others' work on the subject, was acquainted with some of the mesmerist luminaries of his day, and repeatedly referenced the topic in his correspondence.[6] Also like Mayo, Poe regarded mesmerism as a particularly vivid instantiation of a more fundamental process wherein persons are subjected to the control of universal forces. Unlike Mayo, however, Poe radically deviated from the utopianism of most discourses on mesmerism, picturing instead the unsettling implications of the idea that persons are less sovereign entities than puppets of inhuman powers. Rather than optimistically assuming that such powers must serve human ends, Poe concluded that an all-encompassing cosmic energy inevitably troubles human-being by suspending the autonomy and interiority of individual humans. For Poe, mesmerism revealed the self's identity, its putative independence and integrity, to be disturbingly fragile, if not altogether illusory.

Poe here touches on the constitutive paradox of many nineteenth-century metaphysical discourses: the idea that cosmic forces simultaneously empower and evacuate the discrete self (consider Whitman's "One's-Self I Sing," where he sings both of himself as "a simple separate person" and of his incorporation within the impersonal "Life immense in passion, pulse, and power").[7] Never merely academic, such impossibly self-affirming unions were at the heart of the American dialectic between fierce individualism and those suprapersonal "forces" (whether political, economic, or religious) understood to both transcend and subtend it. Early-century discourses of democracy (e.g., Tocqueville) and late-century articulations of industrial capitalism (e.g., Carnegie) may be the most familiar examples, but the underlying logic was ubiquitous.[8] Poe, however—famously antidemocratic, antinationalistic, anticapitalistic—makes clear that you can't have it both ways, can't transcend the self for the sake of the self, can't unify the social, much less the universal, without eliminating (the individuality of) individuals.[9]

In Poe's universe, then, a cosmic force exists, but not in the service of human interests, relegating us to, at best, an ephemeral existence: contingent, never assured, constantly endangered. Indeed, the first principle of Poe's cosmology is that the universe actively erodes that which can only heuristically be called "human," "individual," or "self" and does so in a manner that is neither notional nor deferred but corporeal and immediate. The abstract susceptibility of persons to the forces posited by Mesmer, Emerson, Thoreau, Mayo, Humboldt, and others becomes for Poe a graphic, bodily process through which people are crushed by the cosmically invested things of their everyday worlds.[10] This might imply a macabre conservation of the self—insofar as the process is pictured as an alien, external, or unnatural threat—but Poe's vision is far more unsettling: though whirlpools, houses, and birds serve as harbingers of the self's demise, its destruction is ultimately seeded from within, as what draws together person and thing in an annihilative embrace is a principle of *universal* convergence. Far from being external, this principle is the means of our composition, making it only logical that we—transient assemblages of differences both foreign and familiar—eventually will be buried by the inrushing weight of uncannily similar things.[11] Picturing the simultaneous provocation and erasure of subjectivity attending the encounter between "self" and "world," Poe cancels their Cartesian separation such that the object of our fear becomes indistinguishable from our selves. This is Poe's version of the meta/physics of his age, a perverse yet consistent calculus that unites everything in existence under a universal law that, by definition, eliminates all difference—including the human difference. His cosmology is thus an immanent development within romantic and transcendentalist thought, a universalism that assents to the claim that everything is connected but then adds that it's to our peril, as the strangeness within beckons the darkness without.

Neither parody nor ephemera, this cosmology is the blood, sinew, and spirit of Poe's corpus, whether in the grand, universal sweep of *Eureka* or in the claustrophobic microcosms of his short fiction: what happens to characters' bodies in the latter—obliteration, collapse, internment—mirrors what happens to the idea of selfhood in the former. Critics from Charles O'Donnell to Evan Carton and Joan Dayan have noted this surprising consistency, but they have tended to read into it a psychological obsession, a cultural allegory, or a philosophical precursor to existentialism or poststructuralism.[12] As productive and necessary as these interpretations have been, the rigorous materialism of Poe's cosmology is finally irreducible to them; the psychic, the social, and the linguistic are, for Poe, mere by-products, brief coronal flares of cosmic process—effects, not causes. In this chapter, I thus abandon the hermeneutics of distance or depth to follow Poe in imagining what it would mean to take this claim seriously, to inhabit a universe so strange and inhospitable ("as regards the greater truths," Poe writes in a critical essay, "men often err by seeking them at the bottom than at the top").[13] I read clues that were never hidden—or were hidden by being always in the open. Approaching Poe's cosmology in this way—on its own, a-human terms—offers not only new readings of his texts but also new understandings of how his thought relates to both his age's humanist imaginary and certain latter-day posthumanisms, including Deleuzian univocity, thing theory, and the new materialisms—all of which are concerned, like Poe, with the dynamics among processes, persons, and things. "Substitut[ing] . . . experimentation for interpretation" in the spirit of Deleuze and Guattari, I assemble this motley collection as an apparatus for further thought, as an occasion to think *with* Poe and not only *about* him—thereby resurrecting him, an undead assemblage even in life, from a premature burial.[14]

"MESMERIC REVELATION" (1844), the second of Poe's mesmeric tales, illustrates the essence of Poe's cosmos. The story presents through the visions of a dying mesmerized patient (Vankirk) a picture of the universe in which all phenomena, both physical and spiritual, consist of a common material substrate, an "ultimate, or unparticled matter, [that] not only permeates all things but impels all things—and thus *is* all things within itself."[15] In the tradition of Epicurus and Julíen Offray de La Mettríe, this matter is made the basis of minds as well as bodies, but, following Spinoza, it is even more radically regarded as an extension of "God" ("Mesmeric Revelation," 1034), the *substantial* medium through which all constituents of the interconnected cosmos manifest the universal impulse.[16] Yet contra the theistic affirmations of life and self evident in contemporary mesmeric theory, Vankirk claims that consciously connecting to this universal influence is achievable only in a mesmeric trance or death because these states allow for the supercession of our individual, "rudimental" bodies ("Mesmeric Revelation," 1038). The self's death (or its mesmeric equivalent) is the necessary cost of transitioning from a body capable of capturing only a small, contingent spectrum of the totality of being to a state of absolute identification with it: "in the ultimate, unorganized life [death], the external world reaches the whole body.... In unison with it—the whole body vibrates" ("Mesmeric Revelation," 1037–38); or, as Vankirk states earlier, "when I am entranced the senses of my rudimental life are in abeyance, and *I perceive external things directly,* without organs" ("Mesmeric Revelation," 1037; my emphasis). Vankirk elaborates:

> Organs are contrivances by which the individual is brought into sensible relation with particular classes and forms of matter, to the exclusion of other classes and forms. The organs of man are adapted to his rudimental form, and to that only; his ultimate condition, being unorganized, is of unlimited comprehension.... It is to the absence of idiosyncratic organs, therefore, that we must attribute the nearly unlimited perception of the ultimate life. To rudimental beings, organs are the cages necessary to confine them until fledged. ("Mesmeric Revelation," 1037–38)

As out-of-body, self-abeyant events, both mesmerism and death offer literally "direct" experiences of "uni[ty]" with the cosmos. Though these experiences may appear to reify the self by extending it even beyond the grave, their insistence on the absence of any mediation between the self and "the external world" requires that the self become pure externality, dissolved as a discrete entity into the totality of vibrating, "unparticled matter." We continue to exist in death, that is, only to the extent that "we" are indistinguishable from that which is not "us."

Just before he dies, Vankirk makes clear that the difference between normal, "rudimental" senses and these sensations of the "ultimate life" (or death) is one of degree rather than kind; both bodily register the universe's "vibrations," even if it is only in the latter state that one recognizes what the sensations portend: the conjunction of sensor and sensed in the unity of the "unparticled matter." Whether one is alive or dead, awake or entranced, sensations mark the immediacy of congress and reciprocity of identity between universe and subject. "Mesmeric Revelation" underscores this point by making subject and object indistinguishable in the same moment that self and not-self are collapsed into a common union, in the same space where "*external things* [are perceived] *directly*"—perceived as if by God.

Let us return for a moment to Vankirk's description of this limit experience as one of being a body "without organs," an image that is, of course, central to Deleuze and Guattari, who present the BwO as a pure surface, "the *field of immanence* of desire," and a "plane of consistency or of composition" on (and through) which intensities couple, decouple, circulate, and pass in virtually infinite interplay.[17] This may seem far afield from Poe's concerns, but the connections run surprisingly deep: Vankirk's suggestion that reified distinctions between self and world dissolve with the realization that persons are noncontinuous effects of transbodily sensations is consonant with the arguments of the paired *Capitalism and Schizophrenia* volumes, which first diagnose the fascistic consequences of regimenting society around self-contained subjects and then envision an alternative paradigm of nonorganismal, rhizomatic flows of desire. The admonition in Michel Foucault's preface to *Anti-Oedipus* that we "must rise from our graves" into a new form of life, coupled with the authors' own valuation of the liminal space between life and death, only further underscores the resonances with this body of work and Poe's concern with the body.[18]

Obviously, in Poe's writings, deterritorializations and exhumations rarely end well for the subjects experiencing them, but Deleuze and Guattari, unlike many who conjure endless pleasures in their names, acknowledge that such processes are "not reassuring. . . . [They] can be terrifying, and lead you to your death."[19] Deleuze underscores this promise–threat in his later book on Francis Bacon, which examines the painter's creation of a "zone of indistinction" between his Figures—mere "sensible form[s] related to . . . sensation[s]"—and their background environments: Figures "return to the field in order to be truly dissipated in it, to be effaced on the wall of the closed cosmos, to melt into a molecular texture . . . one stage in a . . . profound becoming-imperceptible in which the Figure disappears."[20] The Figure is thus revealed on the canvas in its true colors—having no origin, existence, or future independent of the environment. And in canceling distinctions of form in this manner, Bacon's paintings point to the "ground" that lies behind both Figure and field: "the rhythmic unity of senses . . . beyond the organism . . . [a] profound and almost unlivable Power"—in other words, the BwO, or what Deleuze elsewhere calls the "pure immanence that . . . is A LIFE" (indefinite, impersonal, inhuman, singular but nonindividual).[21] The point, for both Deleuze and Poe, is that this life not our own is not some static transcendental beyond but the creative immanence by which we—whatever we may be—are made and unmade, become and come undone.

Eureka (1848) is Poe's paean to this life. Combining empiricist enquiry and bodily intuition, the prose poem offers a materialist physics of the spiritual relation between self and universe by showing the interdependence-cum-convergence of the two. In fact, *Eureka* is "Mesmeric Revelation" writ large and as a scientific tract, the expansion of mesmeric principles to their natural, universal limits. And as with his mesmeric tales, Poe regarded *Eureka* as essentially nonfictional, even the culmination of his career: "I have no desire to live since I have done 'Eureka,'" he wrote shortly before his death, "I could accomplish nothing more"—a sentiment matched in an earlier letter that announces, "What I have propounded will (in good time) revolutionize the world of Physical and Metaphysical Science. I say this calmly—but I say it."[22] I must disagree, then, with those critics who regard the text as an elaborate hoax.[23] Despite dedicating it "[to] those who feel rather than to those who think—to the dreamers and those who put faith in dreams as in the only realities," Poe never suggests that the science of *Eureka* is inaccurate.[24] Rather, he argues in the truest romantic spirit that feelings, dreams, and Schellingian "intuitive *leaps*" are the means by which mere scientific "facts" are comprehended as universal "Laws" (*Eureka,* 1264–65).[25]

What are these laws? *Eureka* asserts that the universe began with God's self-differentiation into a diffuse nervous system (i.e., the multiplicity of all phenomena, including the human) and will end in an inevitable collapse into restored unity—"*in the Original Unity of the First Thing lies the Secondary Cause of All Things, with the Germ of their Inevitable Annihilation*" (*Eureka,* 1261)—a process already evident in "Newtonian Gravity" (*Eureka,* 1281).[26] And as the reference to gravity indicates, *Eureka,* like the mesmeric tales, presents susceptibility to "external" influence as the necessary corollary to ubiquitous meta/physical interconnection: "*Every atom . . . attracts every other atom, both of its own and of every other body*" (*Eureka,* 1284).[27] Because of this spiritualized gravity (with which Newton would have been sympathetic), all things of the universe are related not only in their origin but also immediately in the present, in the infinite interactions taking place between every atom in existence—so much so that even the "displace[ment] . . . by . . . the billionth of an inch . . . [of a] speck of dust . . . shakes the Moon in her path . . . causes the Sun to be no longer the Sun, and . . . alters forever the destiny of the multitudinous myriads of stars" (*Eureka,* 1286). Consequently, nothing is independent or autonomous, nothing fixed or stable.

And this includes us. To illustrate the point—that gravity necessitates the eventual convergence of universe and self—Poe rhetorically conflates celestial bodies and human ones, claiming that "stellar bodies" and "bodies on the Earth" mutually attract one another (*Eureka,* 1328). Grammatically collapsing selves and worlds into a single term ("bodies") is significant, then, because it mirrors the universe's own state of progressive implosion. It also demonstrates that we are neither exempt from this process nor wholly distinguishable from it.[28] Indeed, *Eureka* argues, in being affected by bodies not our own, "we" become something other than we supposed, something less self-sovereign because more alienable. Contact with the universe estranges our bodies from our selves.

Although gravity is the most obvious indication of this telos—the return to unity through the revocation of individuality—it is not the only means of apprehending our fate; innate within our consciousness is a latent knowledge of our future destiny. Even more startlingly, we are not alone in the knowledge. In Poe's panpsychic universe, all "creatures" are conscious of their (d)evolution:

> All . . . creatures—*all*—those whom you term animate, as well as those to which you deny life for no better reason than that you do not behold it in operation—all . . . are . . . conscious Intelligences; conscious, first, of a proper identity; conscious, secondly and by faint indeterminate glimpses, of an identity with the Divine Being. . . . Of the two classes of consciousness . . . the former will grow weaker, the latter stronger . . . before these myriads of individual Intelligences become blended . . . into One. . . . Man, for example, ceasing imperceptibly to feel himself Man, will at length attain that awfully triumphant epoch when he shall recognize his existence as that of Jehovah. In the meantime bear in mind that all is Life—Life—Life within life—the less within the greater, and all within the *Spirit Divine.* (*Eureka,* 1358)

As *Eureka*'s conclusion, this passage encapsulates the essay's dominant themes: that the coherent identities of "individual" things (especially "Man") ultimately will be lost in the final reconstitution of God; that this is a process whose effects can be felt in the present; that we are not the only things to feel it. Together, these ideas capture the exceeding strangeness of *Eureka*'s universe. On one hand, it makes "us" continuous with the godhead. But, on the other, it does so only by reducing "us" to a mortal knowledge of our common thingness; in revealing that "inanimate" and "animate" things alike are of a shared, divine substance, the text necessarily eliminates our ontological distinction and, with it, our separate existence.

"The Island of the Fay" (1841), which uses the pantheistic language of "life within life . . . and all within the Spirit Divine" seven years before *Eureka,* provides a haunting image of this process.[29] Centered on the narrator's picturesque journey to an island secluded deep in the mountains, the tale describes a fairy's cyclical migrations from light to shade, finally ending in her death and integration with the waters on which she traveled. This is a fitting image for Poe's conception of the assimilation of individual life into a larger field of being, the inevitable loss of one's person for an obliterative identity with the universe: "do [the Fays] yield up their sweet lives as mankind yield up their own? In dying, do they not rather waste away mournfully; rendering unto God little by little their existence . . . exhausting their substance unto dissolution?" ("Island of the Fay," 604). Putatively individual entities bleeding their shadow lives into a stream—of God, of death—until they are both no more and something more: in Poe's cosmology, we are such ephemeral eddies of cosmic flows. Thus, though the story ostensibly frames a romantic scene of aesthetic enjoyment and philosophical contemplation, the principle behind both the aesthetic and the philosophy is inconsistent with the viewer's (and the reader's) continued existence. Like the Fay, with "more of care and uncertainty . . . and less of ecstatic joy," we must "gr[o]w feebler, and far fainter, and more indistinct . . . and . . . bec[o]me whelmed in a shadow more black," leaving only "the mere ghost of [our] former sel[ves]" ("Island of the Fay," 605). With the fall of "darkness over all things" ("Island of the Fay," 605), we, too, are eclipsed, the human "stain upon the landscape" erased ("Island of the Fay," 600).

In opposition, therefore, to the optimism and anthropocentricism of its contemporaries (Deleuze notes the explicitly Hegelian moments when Whitman "finds in pantheism a reason to inflate his own ego"), *Eureka* and "The Island of the Fay" echo Vankirk in making "our" death—the death of the individual, the death of the human—a precondition of Life.[30] "In order to comprehend what [God] is," Poe notes, "we should have to be God ourselves" (*Eureka*, 1276)—and, for Poe, we can only become God when "we" are no more. Something survives our fatal convergence into the "Material Nihility" of "absolute Unity" (*Eureka*, 1355), but it is not "us"; in Poe's universe, identity can only be born in the moment that difference is buried, when gravity yields the grave.[31]

Closest to *Eureka* in scope and transparency of theme, Poe's three cosmological dialogue-fictions—"The Conversation of Eiros and Charmion" (1839), "The Colloquy of Monos and Una" (1841), and "The Power of Words" (1845)—illustrate this self-annihilative unity even more directly. As in the mesmeric stories, these tales' characters access universal truth only in death. In the first, the deceased

Eiros, now "overburthened with the majesty of all things," recounts the destruction of the world by a comet which had effects even prior to its arrival: "We saw it not as an astronomical phenomenon in the heavens, but as an incubus upon our hearts, and a shadow upon our brains. . . . It was clear that we were already within the influence of the comet; yet we lived. We even felt an unusual elasticity of frame and vivacity of mind."[32] Consonant with the claims in "The Power of Words" that "no act is without infinite result" and that "vibrations [are] indefinitely extended, till [they] g[i]ve impulse to every particle of the earth's air, which thenceforward, *and for ever,* [are] actuated by the one movement," the comet embodies how universal forces impact "us"—even prior to final contact.[33] And as the logical culmination of the closing of the distance between "us" and the universe, the comet's fiery concussion literally fuses "us" with the material world in "a wild lurid light . . . penetrating all things." *Eureka*'s bucolic "flash" of "the tribe of stars . . . into a common embrace" (*Eureka,* 1353) is translated here as a terrifying "species of intense flame . . . [an] all-fervid heat" that ends the "trembl[ing] and shriek[ing]" of "all men" by extinguishing "man" as a discrete category.[34]

Monos, of "The Colloquy of Monos and Una," also dies in a "fiery overthrow" of the Earth, his death, too, testifying to the obliterative conjunction of self and world. After a year interred—his individuality disintegrated, his sense of self lost—he tells his companion, Una, how he became his grave:

> The consciousness of *being* had grown hourly more indistinct, and that of mere *locality* had, in great measure, usurped its position. The idea of entity was becoming merged in that of *place.* The narrow space immediately surrounding what had been the body, was now growing to be the body itself. . . . The sense of being had at length utterly departed, and there reigned in its stead—instead of all things—dominant and perpetual—the autocrats *Place* and *Time.* For *that* which *was not*—for that which had no form—for that which had no thought—for that which had no sentience—for that which was soulless, yet of which matter formed no portion—for all this nothingness, yet for all this immortality, the grave was still a home, and the corrosive hours co-mates.[35]

Transitioning from the sentience of an individual to a sensation of deindividuation, Monos here records the process by which the self's becoming-earth decays human rationality and psychology. Now a hollow coordinate plotted on the axes of "*Place*" and "*Time,*" a diffuse "narrow space" rather than an integrated person, a temporal duration of "*that* which *was not*" rather than a self-present "being," Monos is at last one with his environment. "Soulless" and immaterial, even his name ("Monos," like "Una" in Latin, translates to "one" in Greek) bespeaks his lack of a separate existence. Unlike the self-affirmative harmonies of American transcendentalism, however, it is a "corrosive" unity won only by becoming one with "nothingness"; against the tendency in some posthumanisms to extend our subjectivity into the world, Poe here pictures a self caving in under the weight of its surroundings.

And what happens to Monos in death is only the culmination of a lifelong process; he dies of a "fierce fever" explicitly attributed to the "infect[ion]" of "system" affecting the planet: "in the Earth's dotage," he was "[w]earied at heart with anxieties which had their origin in the general turmoil and decay" ("Colloquy," 612). Monos, in other words, is integrated with the Earth to such an extent that he feels its sickness as his own. We can thus say that Monos does not survive his death so much as he outlives the illusion of his discrete self. What remains is a paradoxical simultaneity of sameness and difference—the "immortality" of "nothingness."

Poe's cosmological tales thus exhibit locally what *Eureka* demonstrates generally: the effect of cosmic integration on our selves, with individual beings absorbed into general Being or, stranger still, becoming. Other stories, however, make even more tangible what *Eureka* only hints at via references to "spiritual shadows" (*Eureka,* 1356), "phantoms of processes completed long in the Past" (*Eureka,* 1321), and "*Memories* that haunt us during our Youth," all of which "sometime pursue us even into our Manhood:—assum[ing] less and less indefinite shapes:—now and then speak[ing] to us with low voices" (*Eureka,* 1357). Indeed, Poe claims, "we walk about . . . encompassed by dim but ever present *Memories* of a Destiny more vast—very distant in the by-gone time, and infinitely awful" (*Eureka,* 1356). These "memories," Poe suggests, are of our past unity, our existence prior to our differentiation from One, and therefore are also necessarily omens of our future destiny. The temporality inherent to a memory that is simultaneously a prophesy speaks to the fact that our ostensibly individual lives are bracketed by an unindividuated mode of being. It also illustrates that even in this life, "we" are not truly independent; "shadows" and "memories" "speak to," "pursue," and "haunt" us in the present, reminding/foretelling us of our irrevocable/inevitable past/future nonexistence, recalling to us—at first faintly and then ever louder—that "we" never are (alone).

Such a scenario should sound familiar: it is one that Poe's tales obsessively rehearse. Time and again, Poe's characters are "haunted" by the knowledge of something "more vast" and "infinitely awful," something they faintly apprehend as a matter of life and death. In the tales, though, the ephemeral whispers of *Eureka* assume material bodies: "shadows" take flesh; "phantoms" become palpable realities; incorporeal thoughts "assume less and less indefinite shapes." Yet these now tactile places and things do the same work as *Eureka*'s fleeting memories: they "speak to us with low voices" about our once and future nihility. Localized instantiations of the fatal effect of other bodies on our own, the myriad familiar-yet-threatening forms in Poe's tales concretize *Eureka*'s cosmic narrative—individual entities subjected to the influence of their environments until a final collapse cancels their differentiation. From cats to comets to teeth, normally innocuous things, now synecdoches of *Eureka*'s cosmic import, bring selves into submission to the world by erasing the distance between the two.

Heidegger's analysis in *Being and Time* of the relation between *Dasein* and objects can help clarify the significance of what Poe is doing in these instances. Exploring the paradox that everyday things (hammers, clocks, etc.) are normally invisible as discrete, ready-to-hand items because of the familiarity and banality of their use ("'things' never show themselves initially by themselves"), Heidegger argues that they "become conspicuous"—are present-to-hand—only when they resist the normative associations attached to them or when, more directly, they defy being used.[36] Whether because damaged, misplaced, or merely "in the way," things demand confrontation only when they are "obstinately" "*un*handy," when their "handiness" or purposiveness for *Dasein* is interrupted (*Being and Time*, 69). What is important about this interruption is that it briefly suspends *Dasein*'s normal being-in-the-world by short-circuiting *Dasein*'s ability to "tak[e] care of things" (*Being and Time,* 71), which, for Heidegger, has deep implications. Because of the irrevocable interconnection between *Dasein* and environment, indeed their mutual definition, if *Dasein*'s being-in-the-world were disrupted in more than a momentary way by the insubordination of things (something Heidegger does not explore), then *Dasein*'s "fundamental structure" (*Being and Time,* 37), or "basic ontological constitution" (*Being and Time,* 181), necessarily would be upended.

Although it may seem odd to reference Heidegger when discussing Poe, both examine how aberrations from the normal uncover that which was previously unseen: the union of self and "*pragmata*" in Being (*Being and Time,* 64). Despite this common starting point, however, the authors arrive at illustratively divergent conclusions. Though they agree that neither self nor world exists prior to its encounter, Poe refuses to grant priority in the intersection to *Dasein,* as Heidegger does.[37] Indeed, Poe's tales invert Heidegger's teleology of the relation between the ontological being of *Dasein* and the ontic being of objects: Poe makes objects significant because of their *canceling* (rather than furtherance or explanation) of human-being. Moreover, for Poe, this resistance by things raises the unsettling possibility that human intelligence is not the only consciousness-of-being extant within the universe, potentially not even the highest (a conceptual impossibility for early Heidegger). Poe presents things as diverse as landscapes, animals, and atoms as sentient and agonistic, consciously thinking of, manipulating, and destroying not only his characters but also the very distinction between object and subject.[38]

In so doing, Poe anticipates both Graham Harman's critique of Heidegger's anthropocentrism and his proffered correction: an object-oriented ontology that accords objects existence and agency beyond our use of them.[39] Unlike Harman, however, Poe does not posit that these things have an inner essence withdrawn from all relation, contending instead—with Whitehead, Latour, and Steven Shaviro—that the line between "us" and "them" is frightfully unclear.[40] As in Bill Brown's "thing theory" and Jane Bennett's "thing-power," Poe's things exist "beyond the grid of intelligibility," "as a recognizable yet illegible remainder or as the entifiable that is unspecifiable."[41] But, and this is the rub, our inability to fix things arises not from abyssal distance but from fearful proximity because we are ourselves monsters, "impure . . . human–nonhuman assemblages" (Bennett), meaning that "inanimate objects" not only "threaten . . . human subjects," they also "constitute . . . them" (Brown).[42] Where Heidegger wants to eschew "theoretically" (*Being and Time,* 68) considering things independently of their "handiness" to *Dasein* as "material for living" (*Being and Time,* 64), Poe wants to demystify the illusion of independence by revealing there to be only one material of life.[43] With the "new" materialisms, Poe imagines "us" to be the product of "cosmic forces assembling and disintegrating to forge more or less enduring patterns that may provisionally exhibit internally coherent, efficacious organization. . . . In this monolithic but multiply tiered ontology, there is no definitive break between sentient and nonsentient entities or between material and spiritual phenomena."[44] For Poe, therefore, difference is never wholly external, never a relation between static identities; it is instead an immanent, self-generative event (Deleuze's "difference in itself") that simultaneously makes and alienates bodies by folding the outside in.[45] The difference *is* (the) within. The supposedly individual self, actually coextensive with the universe, is thus both more (God) and less (an atom) than itself, but always already different from itself: "All existing bodies . . . are . . . mere assemblages of more or fewer differences" (*Eureka,* 1282).[46]

What *Eureka* effects, then, is a transition from digital–synchronic conceptions of self and other to an analogic–diachronic understanding that cancels absolute divisions by illustrating the flux between individuation and identity.[47] It bears repeating, however, that despite the potentially auspicious resonances of this assertion, Poe makes clear that it is neither allegorical (of, say, the relation between citizen and state in U.S. democracy, as W. C. Harris contends) nor anything to which "we" can have a positive relation: "we" can't put it to use, be elevated by it, pray to it, take solace from it; "we" can't even survive it.[48] "We" can only be dis/integrated by it, a fate that is ours even now: "The inevitable catastrophe is at hand" (*Eureka,* 1353).

Rather than asserting their presence only when "unhandy," then, Poe's things attain prominence by forcefully propelling themselves from the indiscernibility of mere materiality to the hypervisibility of people's lives—before all fall back, corporately, to the abyss of indistinction. Indeed, it is this moment of fatal relapse into thingness that makes "individual" forms apprehensible at all—recall that Vankirk is most aware of his body when he senses its imminent loss. Like stars speeding into the oblivion of an event horizon, Poe's characters are brightest just before the black. And this fusion-via-annihilation is Poe's template for all subject–object relations, with individual things standing in as the reminders of "unparticled matter" as such, the common denominator that we cannot bear to acknowledge in ourselves. Corporealizing the universal processes of *Eureka,* the cosmological fictions, and the mesmeric tales, the protean things of Poe's other writings thus do violence to characters by attacking their integrity from within and without; the tales show the often terrifying process of the self becoming, as it were, a thing it-self. Put another way, Poe's tales ask: What happens when Heidegger's hammer hammers us? Or when a raven not only speaks, but speaks in the negative?[49]

Describing its narrator's "destr[uction]" by "a series of mere household events," "The Black Cat" (1843) offers one answer.[50] Ostensibly a perverse temperance narrative about a man's alcohol-induced transformation from loving husband and pet owner into inhuman murderer of his animals and spouse, the story appropriates the destructive power of Eiros's comet for a domestic animal: after the narrator tortures and then hangs one cat, a new cat (or a reincarnation of the old) arrives; and when shortly thereafter it is walled up with the narrator's murdered wife, its "inhuman" "howl" leads police to discover the crime, thereby consigning the narrator "to the hangman" ("Black Cat," 859).

Justice is thus served, but there is more to the story than karmic retribution. There is something about the second cat that is irreducible to the tale's apparent moral, something in excess of the human drama of crime and punishment. Specifically, the cat seems to be a material harbinger of *Eureka*'s vision of the fatal collapse of all things into each other. Like *Eureka*'s "phantoms" and "spiritual shadows," it is described as an "incarnate Night-Mare" and a "vast weight . . . incumbent eternally upon my *heart*!" ("Black Cat," 856). Similarly, the narrator feels "terror . . . horror" and "an absolute *dread* of the beast," which is "not exactly a dread of physical evil," yet one that he is "at a loss how otherwise to define" ("Black Cat," 855). This "chimera" ("Black Cat," 855) of indefinite fear manifests *Eureka*'s claim that we are "haunted" by a subconscious knowledge of our impending and "infinitely awful" convergence with the world. Here, however, the revelation is never fully grasped, the loss of one's individual identity as "Man" a tragedy to be fought rather than a fait accompli: "[For] *a brute beast* to work out for *me*—for me a man, fashioned in the image of the High God—so much of insufferable wo!" ("Black Cat," 856).[51] Refusing to suffer this assault on his divinely ordained manhood, the narrator attempts to reestablish his dominance over the "beast" by "brut[alizing]" it. Of course, the attempt fails, as, according to the story's logic, it must; they are inextricably bound to one another, the image of the narrator awakening to "find the hot breath of *the thing* upon my face" ("Black Cat," 856) a testament to their unity of spirit (i.e., the cat's exhalation is the man's inspiration). Like the white patch resembling a "GALLOWS" on its chest ("Black Cat," 855), the cat's howl thus echoes what "the low voices" of *Eureka* foretell: the end of the narrator's independent life. From this perspective, the titular subject's reincarnation doubles the return of the narrator's repressed, which is less his entombed wife than his once and future union with the supposedly external world.

Perhaps most telling, however, is the narrator's professed motive for torturing his cats in the first place: "the spirit of PERVERSENESS," subsequently described as an "unfathomable longing of the soul *to vex itself*—to offer violence to its own nature—to do wrong for the wrong's sake only" ("Black Cat," 852). What is significant about this description is its equation of harming other things with harming oneself: the soul does "violence to its own nature" by violating something else. However obliquely, the narrator apprehends that there is a "*Law*" that connects him to the animal he harms, which means that murder is inevitably an act of suicide ("Black Cat," 852). Yet far from being an ethical ground, such self-(other)-destruction proves inescapable. As "The Imp of the Perverse" (1845) makes clear, the compulsion to self-obliteration is constitutive of existence, "an innate and primitive principle" that is "antagonistical" to "self-defence."[52] According to the "Imp" and the narrative trajectory of "The Black Cat" (which also identifies this impulse, seventy-five years before Freud, as an "indivisible primary facult[y] . . . of Man" ["Black Cat," 852]), that which most defines the self is paradoxically that which ensures the self's extinction. This is *Eureka* by another name, of course, with its inevitable process of universal annihilation translated into an irresistible drive to self-harm. In both, individual entities rush to their eradication through encounters with what only appear to be their others; once again, the universe's interconnectedness is made manifest through mutual destruction.

"Ligeia" (1838) takes this fatality of unity as its principal theme—and clothes it in human rather than animal form. Given its epigraph, though—which reads, in part, "God is but a great will pervading all things. . . . Man doth not yield himself to . . . death . . . save only through the weakness of his feeble will"—and given the standard interpretation of the story as emblematic of an indomitable will enabling even the survival of death, it may seem counterintuitive to include the tale in a discussion on the death of individuals.[53] The epigraph, however, anticipates almost verbatim *Eureka*'s pantheistic assertion of God "pervading all things," including the "will" of "Man." And as we have seen, such universal interconnection works to cancel the survival of individual selves. Indeed, from the tale's first page, the narrator's existence as an independent entity is endangered, as Ligeia is characterized as "enthralling" ("Ligeia," 310) and "pass[ing] into [his] spirit" ("Ligeia," 314); he even admits, "I was sufficiently aware of her infinite supremacy to resign myself . . . to her guidance" ("Ligeia," 316). Reproducing the structure of the mesmeric séance, the narrator literally loses himself in Ligeia's eyes, his "I" into her eye.[54]

To see how this represents a loss of selves rather than their consummation, let us follow the narrator's example and pause for a moment on (or within) Ligeia's eyes. To him they resemble the "twin stars of Leda" and almost "recall to memory something long forgotten" ("Ligeia," 313). But he is not "able, in the end, to remember" what this "something" is ("how frequently . . . have I felt approaching the full knowledge of their expression . . . yet [it] not quite be mine"), leading him to repeatedly ask "what *was* it?" regarding that "which lay far within the pupils of my beloved" ("Ligeia," 313–14). Groping for an answer, the narrator compares looking into her eyes with viewing "the commonest objects of the universe": "a rapidly growing vine," "a moth," "a butterfly," "a chrysalis," "a stream of running water," "a falling meteor," the "glances of unusually aged people," "books," and "stringed instruments" ("Ligeia," 314). What unifies these things? Some arcane vital energy? Eternal process? In either case, they would mirror Ligeia's indefatigable "will" to survive death. But if this is true, then their relation inexorably depersonalizes Ligeia insofar as her defining characteristic is found to be nothing less than the common nature of the universe itself. Perhaps, then, the association of these disparate things and Ligeia registers a trace of their shared origin and present interconnection; perhaps they are clues to the narrator's "something long forgotten" or to *Eureka*'s "spiritual shadows" and "*Memories* of a Destiny more vast." Perhaps these things, in other words, are the forensic evidence of what *Eureka* describes as God's past self-dispersal and eventual reconstitution.

If so, then Ligeia's eyes, only the most prominent feature of a body that the narrator comprehensively anatomizes, testify neither to personal uniqueness nor to a peculiarly human quality but to continuity with the universe; even their spherical form is patterned tellingly on inhuman "spheres" of a larger scale: stars and cosmoses. These comparisons thus suggest "a *circle* of analogies" ("Ligeia," 314; my emphasis)—a *figural* relation—because they refer in the first instance to a *genealogical* resemblance: her "large and luminous orbs" (a recurrent description) are rhetorically *and* materially related to the heavenly orbs above her ("Ligeia," 314). Again, what this connection reveals is not what makes Ligeia exceptional but what makes her "common" (equivalent to "the commonest objects of the universe"). Or, if she is exceptional, then it is by the degree to which her commonality is evident. Indeed, the tale as a whole denies individuality to its characters by interlineating them with their supposed others: the narrator into Ligeia, Ligeia into the indistinction of a universe of kindred things. As in "The Man That Was Used Up" (1839), what appear to be persons—closed singularities—turn out to be ever-receding series of prostheses.

Ligeia, uncannily attuned to this process, pens a macabre poem about it from her deathbed. Describing the fate of humans in the language of a stage "tragedy" titled "Man," her verses depict a spectral orchestra playing "the music of the spheres"; characterize humanity as "Mimes" and "puppets" controlled by "vast formless things," "chas[ing]" but never "seiz[ing]" the "Phantom" they seek; and conceive of life as "a circle that ever returneth in / To the self-same spot," ostensibly the self's return to nothingness via the "Conqueror Worm." With "the soul of the plot" being "Madness," "Horror," and "Sin" ("Ligeia," 318–19), the poem thus calculates the "human" cost of *Eureka*'s reversion of selves back to One: what *Eureka* characterizes as a painless fading of self into universe, the poem presents as a painful farce on the impotence and impermanence of man. The motion is the same, the emotion differs.

Even Ligeia's "return" from death, which apparently overthrows the Conqueror Worm's reign, is not an assertion of the durability of individuality but an exemplification of the forces obliterating it. Amid the "hideous and uneasy animation" of a tapestry ambiguously moved by spirit or wind, and the "phantasmagoric influences of the chamber itself" ("Ligeia," 322, 323–24), the narrator's new wife, Rowena, dies, and by degrees her body is transformed into that of Ligeia, who rises again to life. But what does it mean to say that Ligeia has returned? Who or what is Ligeia? At the beginning of the tale, the narrator confesses knowing neither how they met nor her "paternal name" ("Ligeia," 311). And aside from abstract physical descriptions that veer to comparisons with inanimate things, Ligeia is characterized only as having an "immense" amount of "learning" in "transcendentalism" ("Ligeia," 316), a "gigantic volition" ("Ligeia," 315), and, when nearing death, a "wild desire for life,—for life—*but* for life" ("Ligeia," 317). Indeed, the only words we hear her speak ("Man doth not yield . . . to the angels, *nor unto death utterly,* save only through the weakness of his feeble will") echo the tale's epigraph, which is attributed to someone else entirely ("Ligeia," 319). In every way, then, "Ligeia" deindividualizes Ligeia by denying her qualities other than those of the universe as a whole. Embodying *Eureka*'s "Divine Volition" and "Life—Life—Life within life," Ligeia is a transcendental abstraction made concrete. What survives death is not a person but a principle.[55] What returns from the grave is Life itself.

If "Ligeia" portends our ontological dis/integration by making Ligeia a lodestar of cosmic process, "The Fall of the House of Usher" (1839), written only a year later, extends this office to even more things, picturing yet more ways in which we are "puppets" of our surroundings.

Indeed, the story offers numerous synecdochic examples of the concrete forms universal forces can assume, the most prominent of which is, of course, the House itself. Upon first arriving, the narrator immediately registers the "unnerv[ing]" lifelikeness of the place: "The melancholy House of Usher" twice is said to have "vacant eye-like windows"; indoors, too, things have their own "atmosphere of sorrow" and "irredeemable gloom" that "h[angs] over and pervade[s] all."[56] The affective resonance of the location leads the narrator to question, in a manner reminiscent of Ligeia's husband, "What was it . . . that so unnerved me in the contemplation of the House of Usher?" ("Usher," 397). Though he does not arrive at an answer, he entertains a "strange fancy" that the environs are somehow communally alive in their shared decay: "About the whole mansion and domain there hung an atmosphere peculiar to themselves . . . an atmosphere which had no affinity with the air of heaven, but which had reeked up from the decayed trees, and the gray wall, and the silent tarn—a pestilent and mystic vapor" ("Usher," 399–400).

Even more unnerving, the house's "atmosphere" quickly becomes the narrator's own: "With the first glimpse of the building, a sense of insufferable gloom pervade[s]" his "spirit," resulting in an "utter depression of soul" analogous to "the bitter lapse into everyday life" of waking from an opium reverie ("Usher," 397). Subjected to "the vivid force of the sensations which oppressed [him]," and unable to reestablish his autonomy by "shaking off from [his] spirit what *must* have been a dream" ("Usher," 399), the narrator is forced to confess that "there *are* combinations of . . . natural objects which have the power of thus affecting us" ("Usher," 398). Though "analysis of this power lies . . . beyond [his] depth" ("Usher," 398), the fact that the narrator feels it at all suggests his permeability to the outside. Here, it seems, affects reside in things and then possess persons, not the other way around.

All of this could still be discounted as the narrator's "shadowy fancies" ("Usher," 398) were it not for what follows. Literalizing the indistinguishability of building and inhabitants inherent to the "equivocal appellation of the 'House of Usher,'" the story documents the effects of the "influence" of structure on lineage over "the long lapse of centuries," an imbrication made all the more incestuous because the "very ancient" family tree never "put forth . . . any enduring branch" that lived beyond the mansion's walls ("Usher," 399). Suffering equally from lines that are absences (the branchless family tree; the "fissure" reaching from the top of the house's roof to its foundation), the house and its eponymous owners are connected by a common rupture that foreshadows their mutual collapse ("Usher," 400). And Roderick Usher, last of the family line, knows it:

> He was enchained by certain superstitious impressions in regard to the dwelling which he tenanted, and whence, for many years, he had never ventured forth—in regard to an influence . . . which some peculiarities in the mere form and substance of his family mansion, had, by dint of long sufferance . . . obtained over his spirit—an effect which the *physique* of the gray walls and turrets, and of the dim tarn into which they all looked down, had, at length, brought upon the *morale* of his existence.[57] ("Usher," 403)

Baldly presenting the degree of "influence" places have over persons, this passage stresses Roderick's consciousness of the causal link between "the mere form and substance of his family mansion" and "the *morale* of his existence." He knows that his body and "spirit" are only conditional, ephemeral "effect[s] . . . [of] the *physique* of the gray walls and turrets." It would appear that the house is the real body and he the "shado[w]"; it is substantial where he, a mere reflection, is not.

The house's "superstitious force" ("Usher," 403) is subsequently revealed to originate in its vitality, even consciousness, as Usher's general belief in "the sentience of all vegetable things . . . assume[s] a more daring character, and trespasse[s] . . . upon the kingdom of inorganization":

> The conditions of the sentience had been . . . fulfilled in the . . . collocation of these stones . . . as well as in that of the many *fungi* which overspread them . . . above all, in the long undisturbed endurance of this arrangement, and in its reduplication in the still waters of the tarn. . . . The evidence of the sentience . . . was to be seen . . . in the gradual yet certain condensation of an atmosphere of their own about the waters and the walls. The result was discoverable . . . in that silent, yet importunate and terrible influence

> which for centuries had moulded the destinies of his family, and which made *him* . . . what he was. ("Usher," 408)

An emblem of the ubiquity of life and consciousness, the panpsychic house gives form to the nebulous universals presented in *Eureka,* concretizing the essay's broad assertion of the deindividualizing effect of material unity. Fusing stones, fungi, trees, tarn, temporal duration ("centuries"), and observer, the scene demonstrates *Eureka*'s "collocation" of seemingly discrete things, including persons, into one (in)organic whole. The Usher family, then, rather than freely tenanting the building, are themselves possessed; though Roderick is the presumed master of the ancestral manor, it is sovereign over him: the moldering house "mould[s]" the family's destiny and "ma[kes]" Roderick "what he [is]."

Tellingly, he senses rather than knows this truth. Roderick's defining characteristic, and the reason he calls the narrator to visit him, is a hysterical form of oversensitivity, "an excessive nervous agitation":

> The nature of his malady . . . was . . . a constitutional and family evil . . . [a] nervous affection. . . . It displayed itself in a host of unnatural sensations. . . . He suffered much from a morbid acuteness of the senses; the most insipid food was alone endurable; he could wear only garments of certain texture; the odors of all flowers were oppressive; his eyes were tortured by even a faint light; and there were but peculiar sounds, and these from stringed instruments, which did not inspire him with horror. ("Usher," 402–3)

As is true of so many of Poe's characters, Roderick is an unsustainable cacophony of "unnatural sensations," a discordant, refracted reflex of his surroundings.[58] He is himself an instrument, overstrung to the point of breaking and ruthlessly played on by all around him. And the result of this sensual connection is not an exaltation of the subject and its powers but an attenuation of him and them, the production of a state of being that threatens Roderick's life by overwhelming it with context. Roderick himself anticipates the inescapable result of living in constant "terror" of "even the most trivial" thing: "I shall perish . . . I *must* perish" ("Usher," 403).

But what exactly is threatening here? Whence the "horror" and "terror"? Roderick, at least, feels that his connection with the estate threatens his being, and the story proves him correct; Roderick dies at tale's end. But why? That other things' life and sentience should endanger our own is perhaps confusing, especially given the generally positive tenor of panpsychic discourse at the time (as evident in Poe's citations elsewhere of Percy Bysshe Shelley's "The Sensitive Plant" and in Roderick's own library).[59] But insofar as human life is predicated on an assumption of preeminence and exclusivity, it makes sense that "we" would be disturbed if our defining characteristics were distributed. Moreover, in addition to lessening our ontological distinctiveness, Poe's panpsychism also calls into question "our" autonomy: it is no longer clear in what sense "we" can be said to exist as independent beings once our physical, mental, and spiritual lives are revealed to be extensions of sentient environments. In Poe's universe, sovereignty is a zero-sum principle, pertaining either to a universe or to a self, but never to both. The house's consciousness thus upsets the integrity of Roderick's mind by becoming its center, a fact indicated by the ready slippage between Roderick's self-consciousness of his "malady" and his obsessive consciousness of the consciousnesses of other things. For Poe, the unity of the environment, its "peculiar atmosphere" "of [its] own," comes at the cost of the individuality of its component parts. With unity there is scarcity, with singularity there is indistinction. For the Ushers, therefore, there is extinction.

The canonical interpretation of this conjunction of house and family is that it somehow manifests the psychological pathologies of Roderick, the narrator, Poe, or all three.[60] This answer is unsatisfying, however, inasmuch as it refuses to countenance what the tale says things do and shows things doing; it bows to less strange because more human conventions, the very conventions the text undermines. Roderick may be insane, and he may infect the narrator with his insanity, but this is an effect rather than a cause of the fantastical phenomena the story presents. As the narrator only partially realizes, "much, if not all of what [he] felt, was due to the bewildering influence of the gloomy furniture of the room" ("Usher," 411). Such a belief, like that regarding the "sentience of all vegetable things," is not itself insane in Poe's universe, but it may cause insanity, one symptom of the self-estrangement that contact with a vital world produces. Alternatively, insanity or a "morbid acuteness of the senses" may actually enable apprehension of the normally unperceived nature of the things about "us," thus serving the same function as other altered states in Poe's fiction (mesmeric trances, opium reveries, swoons, death).[61] Either way, Poe demonstrates that self and world are defined not by their distance but by the frisson of their annihilative convergence.

My reading has thus far ignored what much criticism on "Usher" has foregrounded: Roderick's relationship with Madeline. The pairing is significant, but not for the reasons often cited. Revealed to be twins of "striking similitude," the resemblance is discovered by the narrator only following Madeline's death ("Usher," 410). In fact, aside from a spectral appearance made at the beginning of the story in the exact moment Roderick speaks of her imminent "decease," Madeline is absent from most of the tale (both physiologically—she is cataleptic—and spatially, as she is confined to bed, out of view), becoming truly present only in "death." Once a "corpse," however, once a thing like Ligeia rather than a person, her presence is felt everywhere. Roderick, especially, senses his sister's return. Owing in part to the "sympathies of a scarcely intelligible nature [that] had always existed between them" ("Usher," 410), and in part to his acute hypersensitivity, Roderick hears "the low and indefinite sounds" ("Usher," 411) of Madeline's efforts at escaping her tomb for days. At the last moments of her approach, his body even begins to register Madeline's proximity, as he starts "rock[ing] from side to side with a gentle yet constant and uniform sway" ("Usher," 415). Like the sympathy between proximate bodies in *Eureka,* Madeline's advance toward her brother changes "the motion of his body" ("Usher," 415). And like *Eureka*'s final, inevitable collapse of differentiated matter into a "common embrace" of indistinction, Madeline's "[falling] heavily inward upon the person of her brother" brings both to the ground as "corpse[s]" ("Usher," 416–17). The subsequent collapse of the house only punctuates the absoluteness of the convergence. Resonating with the language of universal energies, a "life-like" "whirlwind," a "gaseous exhalation which hung about and enshrouded the mansion," and "electrical phenomena" destroy the property, submerging it below the opaque waters of the tarn ("Usher," 412–13).[62] In a significant anticipation of *Eureka,* only the "blood-red moon" remains over the place where persons once were ("Usher," 417).

The common thread throughout the story, then, is the inwardness of identity. Roderick's nervous sensitivity to the mortal influence of a sentient environment, Madeline's "inward" fall onto her brother, and the house's collapse all trope the inexorable return to unity of spiritually interconnected matter ("Usher," 416). In a tale obsessed with linearity (genealogical lines, poetic lines, sight lines, structural lines), the narrative traces how discrete points finally coalesce back into a single point. Consequently, the line that divides self from world, the ontological barrier visualized in the written "I," is shown to be incapable of withstanding the pressure of what lies without. The Ushers and the House of Usher necessarily fall together because the logic of "The Fall of the House of Usher" is that identity is borne out in collapse and unity is expressed in erasure. The tale thus exemplifies that, in Poe's universe, *all* unions are incestuous and issue in death.

Most terrifying of all, Poe insists that in reading the Ushers' fate we share it. His aesthetic theory, consistent with his cosmogony, is predicated on the idea that readers are, like Roderick, effects of their "environments," subjects of "external" sensations both constitutive and destabilizing: "During the hour of perusal the soul of the reader is at the writer's control."[63] The rationale of this claim, the same as that of mesmerism, is that readerly sensations register a spiritual event: the loss of the soul's independence via its integration with the putative outside, including those "glories beyond the grave" of which art and world give us only "brief and indeterminate glimpses."[64] Reading, that is, reveals one's meta/physical continuity with the "shadows" of "the spirit's outer world," a fact best embodied in the way that characters' supernal "fanc[ies] ... induce ... similar one[s] on the part of the reader."[65] Accordingly, we are meant to *feel* it when the narrator of "The House of Usher" senses "creeping upon [him], by slow yet certain degrees, the wild influences of [Roderick's] own fantastic yet impressive superstitions" ("Usher," 411). We, too, are supposed to be "[o]verpowered by an intense sentiment of horror, unaccountable yet unendurable" ("Usher," 411); in sympathy, our own bodies should shake with the "irrepressible tremor" that gradually "pervaded [the narrator's] frame" ("Usher," 411). As he is "infected" by Roderick's "instinctive spirit" of "extreme terror" ("Usher," 411), so, in turn, are we. And the same holds true across Poe's tales: characters' depersonalizing sensations affect/effect our own.[66]

This isn't accidental: if we believe Poe, he meticulously designs his texts to achieve such ends, building his poems and tales with "the machinery" best fitted to achieve the "proposed effect."[67] Indeed, in "The Philosophy of Composition" (1846), "The Poetic Principle" (1850), the Hawthorne reviews (1842, 1847), and elsewhere, Poe argues that texts are engines for the production of effects. The entire gambit of "The Philosophy of Composition" is that it reveals, à la "Maezel's Chess-Player," the "wheels and pinions—the tackle for scene-shifting—the step ladders and demon traps" by which Poe constructed "The Raven."[68] We learn, for instance, that it, like "all the rest" of his texts, was "written backwards," with the intended effect coming first and dictating all the supposedly causal particulars of plot and tone: "in the true construction, the cause . . . is absolutely brought about by the effect."[69] According to Poe, this inversion is essential because "the unity or totality *of effect*" is the law of literature.[70]

It is also, uncoincidentally, the law of Poe's universe. Recall *Eureka*'s assertion that the universe was effected by God's dispersal into a diffuse yet unified (and progressively contracting) nervous system, an idea captured in Poe's polyvalent claim that "the Universe is a plot of God."[71] From this perspective, the cosmos is a narrative effect "exactly adapted to its purposes" (*Eureka,* 1300), so that not even a "speck of dust" can be moved without affecting the entire system. The same is true of Poe's ideal artistic plot, defined as "*that from which no component atom can be removed, and in which none of the component atoms can be displaced, without ruin to the whole.*"[72] What is only aspirational for authors, however, is already achieved in the universe: the "absolute reciprocity of adaptation" between cause and effect, "in which we are never able to determine whether any one point depends upon or sustains any one other."[73]

Does this mean that Poe positions himself as God over text and reader alike? Yes and no. On one hand, he does emphasize "the magnetism" and "necromancy of [the author's] words," words that cannot "do otherwise than exhort a response" and "compel [readers] to feel."[74] He even uses the loaded language of the "clay" of the artistic work (and presumably of the reader) being "the slave of the artist," a figuration that later will be literalized in Charles Chesnutt's "Lonesome Ben."[75] On the other hand, though, Poe frequently suggests that the artist, too, is "forced" to bow to the inevitable "progress of the construction"; in fact, "the Poet" is defined by his or her greater *susceptibility* to "the simple elements which *induce* . . . the true poetical effect."[76] In this conception, art, like the universe, "proceed[s], step by step, to its completion with the precision and rigid consequence of a mathematical problem."[77] It is an impersonal, inhuman process "founded in Nature," meaning that authors themselves are not finally causes but effects.[78] They are at most assembly mechanisms because, as Klaus Benesch has argued, Poe sees art not as being created sui generis but as mirroring life in being compositional, achieving its originality by effecting "unusual combinations" of already existing things: "as often happens analogously in physical chemistry, . . . the admixture of two elements results in something" previously unseen.[79] The key to Poe's cosmology is that "this chemistry of the intellect" is not simply analogical; it signals the common origin, current imbrication, and future unity of self, world, and text.[80] It suggests, in the words of Paul Valéry, Poe's "very strict and deeply alluring doctrine, in which a kind of mathematics and a kind of mysticism bec[o]me one."[81]

Poe is God, then, but only to the extent that he claims we all are (in the incoherence of our individual identities), and only in the Spinozist terms outlined by *Eureka,* where God is immanent, not transcendental, a dispersed effect of the system rather than a foreign sovereign. Far from consolidating our position atop the chain of being, such an understanding destabilizes "us," revealing our selves to be not our own, our lives but evanescent episodes within an eternal, impersonal flux. In so doing, it resonates, as we have seen, with much recent posthumanist theory that redistributes agency, sentience, and value beyond our inherited anthropocentrisms. For Poe, however, the recognition of oneself in the world—and of the world in oneself—is gothic rather than utopic, something to be lived, if at all, by the dead. Neither triumphantly nor redemptively posthuman, Poe's version of flat ontology insists on the overcoming of the subject by the universe it uncannily resembles—such that the "I," too, becomes a thing. His closed, deterministic, teleological cosmos—darkly beautiful in its crystalline perfection—defies our attempts to humanize it, to conscript it to our ends, but at the cost of our very existence.[82]

There is thus no room in this vision for a positive ethics (for Poe, the cardinal sin of art is the "inculcat[ion of] a moral," and he broadly rejects "faith in human perfectibility. . . . Human exertion will have no appreciable effect upon humanity").[83] Even if we were to recognize, with the narrator of "The Black Cat," that harming others injures our own estranged being, we would be helpless to act otherwise because the universe's contraction requires the destructive conjunction of ostensibly discrete bodies (atomic, human, planetary). The Many must be sacrificed at the altar of the One. This fatally consistent working out of transcendentalism's logic is both the power and the limit of Poe's cosmology. It may denaturalize the socially contrived inequalities of slavery, for example, but only by revealing that we are all strange hybrids of master and slave. Rather than offering an alternative to what Terence Whalen calls Poe's "average racism," it can only universalize it, making the human race the impossible, untenable average of everything and nothing.[84] Representing an inassimilable negativity *within* the twinned developmental histories of the human and the posthuman, Poe can only offer a meta/physics of universal relativity and, consequently, a different way of thinking about "us."

This is much, but it is not all, and in chapters 3–5, I explore an African diasporic cosmology that disrupts the autonomy and priority of the bounded human subject without necessitating an apocalypse. Here, self and cosmos are interlineated but not coincident, their relation open-ended rather than pre-scripted. Consequently, this cosmos can countenance a world in which "we" are transformed without being extinguished. I turn first, however, to another totalizing vision, Henry Adams's isomorphic histories of himself and his world, which foreclose such a possibility by announcing the imminent collapse of each. Even less human than Poe's "nebular cosmogony" of eternal "re-juvenescence," Adams's entropic universe reverses Poe's plot by moving from unity to dissipation—with no hope of return, thus anticipating (and occasionally preemptively interrogating) much current work in the posthumanist implications of chaos and deep time.[85]

2

HENRY ADAMS'S HALF-LIFE
The Science of Autobiography

If we now say that irony constituted the substance of his existence (that is, to be sure, a contradiction, but it is supposed to be that), and if we further postulate that irony is a negative concept, it is easy to see how difficult it becomes to fix the picture of him.
—Søren Kierkegaard, *The Concept of Irony with Continual Reference to Socrates*

THE EDUCATION OF HENRY ADAMS is a remarkably uninhabited autobiography. Written exclusively in the third person by a narrator who repeatedly refers to "Henry Adams" as "passive," "submissive," and "a helpless victim" before the "forces" of the universe, *The Education* advances a decidedly attenuated model of author and subject.[1] Indeed, one of the most characteristic gestures of *The Education* is its imagination of not only the quiescence-unto-obsolescence of "Adams" but, relatedly, of his extinction altogether. Having cultivated "a singular sympathy for death" over the course of his life (*Education,* 501), Adams's education culminates in the recognition that he is, in fact, already "dead" (*Education,* 320). Suffice it to say, the book has a vexed relationship with its self.

It is true that the first half of *The Education* dutifully records many biographical details of Adams's life: memories of his grandfather, John Quincy Adams; time spent as his diplomat-father's attaché in London during the U.S. Civil War; attending the 1893 Chicago Exposition. But rather than being integrated within an evolving bildungsroman, or even accreting to offer a mosaic of perspectives on a coherent self, these episodes seem strangely disarticulated from their subject, picturing something simultaneously more and less than a self. Like a series of circumstances given an arbitrary, ill-suited name, they seem only casually or accidentally related to "Adams." Indeed, "Adams" seems more the inadvertent effect occasioned by the contingencies that the text documents than their reason, center, or source; he proves, in the final analysis, an insufficient justification of their occurrence. Facts, events, environments, even mere things—all take precedence over eponym. Put another way, the subject—the material—of *The Education* is in tension with its titular subject, Henry Adams.

The Education, moreover, openly dramatizes this tension by increasingly losing its sense of self; after numerous earlier instances of impersonal detachment toward ostensibly personal details, the second half of *The Education* divorces itself almost completely from Adams's biography. Accomplished through what appears to be a genre shift, this withdrawal cedes "personal" narrative to abstract formulation, "subjective" memoir to the objective historiographic calculations of *The Education*'s "Dynamic Theory of History." I say "appears" because this shift—one of degree rather than kind—represents the culmination of the evacuated form of the book as a whole, an acceleration of the text's fundamental law.[2] Because Adams's person is only tenuously present to begin with—and present at all only because of its association with other things—his eclipsing by suprapersonal matters marks not a reversal but a consummation. This is not to say, then, that Adams is nowhere to be found in his text; it is only to notice that the text does not find Adams's self to be an end in itself, or even something particularly worth dwelling on. Consistently prioritizing things other than the "bio," Adams's autobiography makes him secondary to his own story.

THE QUESTION of how much of Henry Adams is in *The Education of Henry Adams* has preoccupied critics since its posthumous publication in 1918. In fact, the dominant critical discourse for the text arguably was set in 1919, when T. S. Eliot wrote that "it is doubtful whether the book ought to be called an autobiography, for there is too little of the author in it."[3] Yet despite the consistency with which subsequent readings have inherited Eliot's view, surprisingly few have considered the explanation the text itself provides for Adams's precarious status. Critics, that is, have largely ignored or disavowed the relationship between the ostensibly biographical elements of *The Education,* on one hand, and the text's parallel articulation of a dynamic theory of history, on the other.

The Education, however, irrevocably binds its self to its theory. Indeed it must, because the dynamic theory, a deterministic worldview intimately engaged with the science and historiography of its period, "reduc[es]" "all history, terrestrial or cosmic, mechanical or intellectual" to a universal "formula" of physics (*Education,* 489), making both the human and the individual self nothing more than dependent effects of nonhuman "forces":

> A dynamic theory . . . takes for granted that the forces of nature capture man. The sum of force attracts; the feeble atom or molecule called man is attracted; he suffers education or growth; he is the sum of the forces that attract him; his body and his thought are alike their product; the movement of the forces controls the progress of his mind, since he can know nothing but the motions which impinge on his senses, whose sum makes education. (*Education,* 474)

> Science had proved that forces, sensible and occult, physical and metaphysical, simple and complex, surround, traverse, vibrate, rotate, propel, attract without stop; that man's senses are conscious of few, and only in a partial degree; but that, from the beginning of organic existence, his consciousness has been induced, expanded, trained in the lines of his sensitiveness; and that the rise of his faculties from a lower power to a higher, or from a narrower to a wider field, may be due to . . . outside force or forces. (*Education,* 487)

The vertiginous language of these passages reproduces the disorientation it describes: the dizzying realization of humanity's susceptibility to infinite externality. Most unsettling of all, however, is the recognition that this externality is not truly external; as in Poe's cosmology, the dynamic theory contends that the very essence of human-being—physiological, cognitive, social—is, and has always been, governed by "outside force or forces," whether they have been "called God or Nature" (*Education,* 487). Consequently, we are not merely "capture[d]" by "the forces of nature," we are their "sum," composed—like "the surface foliage of a vegetable" (*Education,* 475)—by powers irreducible to "our" existence. It's not that "we" don't exist; it's that "we" exist outside of our selves.

Given this revelation, Adams's goal is to find the exact "formula" that would explain human history *and* "satisfy the conditions of the stellar universe" (*Education,* 376). Breathtaking in its generality, the attempt leads Adams to announce that "the historian's business was to follow the track of... energy... as a mathematical problem of influence on human progress" (*Education,* 388–89). *The Education* does exactly that, defining the "phases" of history (*Education,* 342)—from "prehistoric man" (*Education,* 493) to the Virgin to the Dynamo—through differential equations of force. And as hard as it is to fathom, "force" is not metaphorical here; Adams means it to denote a literal, physical law. Unlike with Poe, however, there is nothing behind this law, no prospect of an eventual (albeit obliterative) fusion with God. There's only brute physicality, of which "spirit" is a mere by-product.

The attempt to correlate universal and human history spans Adams's oeuvre, culminating in *The Degradation of the Democratic Dogma,* a posthumous collection of three of Adams's scientific historiographies.[4] The first of these is *The Tendency of History,* Adams's 1894 presidential message to the American Historical Association (delivered, like *The Education,* in absentia), which advocates for a "science of history" that would "fix with mathematical certainty the path which human society has got to follow." The second is "The Rule of Phase Applied to History" (1909), an application of J. Willard Gibbs's chemical rule of phases to the identification of distinct periods or "equilibrium[s]" of civilization. And the third, *A Letter to American Teachers of History* (1910), apocalyptically refers society's present "decrepitude" to the workings of entropy, the "Law of Dissipation."[5] Although these texts do not make identical claims (becoming, as John Conder rightly notes, progressively more pessimistic over time), they all postulate the determinative power of nonhuman forces over human affairs.[6] All three, moreover, apply scientific methods to the study of history, thereby "rais[ing] history to the rank of science" (*Tendency,* 128) and leaving "the future of... History... in the hands of the physicists" ("Rule," 283). *The Education*'s subjection of its author to the dynamic theory thus participates in a project long explored by Adams, one of "converting metaphysics into a branch of physics" (*Letter,* 196), making even "the soul . . . an energy of thermodynamics" (*Letter,* 197).

In this chapter, I consider such meta/physical determinism in its historical and philosophical contexts, examining how it not only allows but necessitates that Adams be out of phase with his own life. Contrary to the scholarly consensus, I argue that it is Adams's science—not his biography—that provides the best explanation for his impoverished status in *The Education*; to attribute it to either stylized detachment or unconscious motivation (both equally "self-centered" causes) is to countermand the ontological imperative that *The Education* embodies: the disintegrative effect of cosmic powers on persons and their stories. Inhumanly consistent, the dynamic theory precludes the possibility of a fully realized or rounded Adams. Instead, inevitably, we are given an "Adams" more atom than man: "[Adams] was . . . a conscious ball of vibrating motions, traversed in every direction by infinite lines of rotation or vibration . . . a centre of supersensual chaos" (*Education,* 460)—where chaos means not the absence of deterministic law but the impossibility of its reconciliation with our self-serving orders ("chaos was the law of nature; Order was the dream of man" [*Education,* 451]).[7]

Thus, rather than dismissing Adams's cosmology as psychological compulsion, ironic performance, or pseudoscientific error, disqualifications that allow us, in the words of Isabelle Stengers, to "withdraw from [our] obligations" to it, I explore the challenge it poses to both the popular evolutionism of Adams's age and the generative mobilizations of chaos and deep time in our own.[8] Adams's scientific historiographies and their equivalents in *The Education* repudiate the mutually informing concepts of the ontologically superior human and the fully agential individual, parallel discourses of upward mobility that found expression in the countless programs of social and personal perfectibility of Adams's era. Before the catastrophic forces of the universe, manifest destiny, scientific management, and social engineering necessarily dissolve. But Adams's theories also, and for the same reasons, resist recruitment into affirmative posthumanisms. Although they anticipate recent engagements with inorganic evolutionism, thermodynamic potentiality, nonhuman temporalities, and speculative realism, their determinism forecloses the possibility of rehabilitating them for recuperative ends. For Adams, chaos is not liberating; it is annihilating. What results is a literal post-human-ism: evacuated of our selves by the cosmos, "we" cannot recohere to enjoy our identity with it. As in Poe's cosmology, but by diffusion rather than condensation, totalization here equals extinction. And if the cost of this antianthropocentrism is ultimately too high, it nevertheless demands a hearing, as it haunts any attempt to derive a livable "us" out of a universe not our own. Thus, as it is for Adams, so will *The Education* be for me an occasion to investigate things beyond (but also within) its self.

THE SIGNIFICANCE of Adams's dynamic theory becomes evident only when examined in the light of what *The Education* names as its intellectual context. In addition to Copernicus, Galileo, Bacon, Descartes, Leibniz, and Newton, *The Education* mentions such nineteenth- and early-twentieth-century scientific luminaries as Michael Faraday, James Clerk Maxwell, Charles Darwin, William Thomson (Lord Kelvin), J. Willard Gibbs, J. B. Stallo, Karl Pearson, Jules Henri Poincaré, Ernst Mach, Arthur Balfour, and Marie Curie. Similarly, Adams's *Degradation* essays are self-avowedly modeled not only on the science of Gibbs and Kelvin but also on the historiographies of Robert Jacques Turgot, Auguste Comte, and Henry Buckle.[9] Though they share little else in common, Adams cites the work of each as a precedent for bringing human existence under the aegis of universal, scientifically apprehensible laws.[10] In so doing, he places himself within a genealogical narrative of modern science, understanding his work as furthering the age's signal assumption that "the forces of nature capture man."

Adams finds the most telling evidence for this assumption in the proliferation of evolutionary literature, both academic and popular, following the publication of Darwin's *The Origin of Species* (1859) and *The Descent of Man* (1871). If not the first (one can think of Buffon's and Lamarck's theories in the eighteenth century), Darwinian evolution was certainly the most widely circulated of mid- to late-nineteenth-century discourses to submit human life to the processes of "natural" law. Rather than remaining a phenomenon apart, the human in evolutionary theory was considered an entity subject, like any other, to nonhuman processes.

Despite this epistemic shift, however, many evolutionist treatises were characterized by the same anthropocentric optimism that typified their predecessors. While scripturally based attacks were leveled against evolutionism because of its untenable denigration of humankind—recall Bishop Wilberforce's famously indignant rejection of the idea that humans were descended from apes—the most prominent evolutionist discourses from 1870 to 1900, especially in America, triumphantly announced the naturalization of the "Supremacy of Mankind over the Earth."[11] As Adams himself notes in *The Tendency,* America added to evolution "the cheerful optimism which gave to Darwin's conclusions the charm of a possible human perfectibility" (130). "Society naturally and instinctively adopted the view that Evolution must be upward" (153), Adams adds in *A Letter,* referring to the work of such popularizers as Ernst Haeckel, Herbert Spencer, and John Fiske, whose theories arguably met with greatest favor in America.[12] If not without controversy, then, evolution—quickly condensed to Spencer's "survival of the fittest"—merely changed the terms in which the ascendancy of the modern human was articulated; though humanity's exceptionality was no longer assured by divine ordination, its position atop creation was yet secured by the workings of natural law.[13]

Inevitably, the notion that humanity represented an evolutionary apex was complicated by protoanthropological accounts of the development of *different* human civilizations. Yet these, too, were soon conscripted into a narrative of advance. Building on a long tradition of idealistic, anthropocentric historiographies that assumed cultural if not biological evolution—Hegel's dialectics of Spirit; Marx's historical materialism; Turgot's successively enlightened economic stages; Comte's positivistic ascension of society and the human mind toward ever-greater truth—Darwinist historians of the late nineteenth century explicitly transferred concepts of natural selection to the rise and fall of various human societies.[14] Prominent examples include Edward Burnett Tylor's and Lewis Henry Morgan's patronizingly racist accounts of the linear "advancement of civilization" as well as Haeckel's zoologically derived biogenic theory of race development.[15] Contributing to what Agamben has identified as the reification of an exclusionary notion of the human by the emerging human sciences of the nineteenth century, these theorists assumed that Western civilization represented either the penultimate stage or the actual pinnacle of biohistorical development.[16] Significantly, then, the (white) human in these discourses becomes a notation for both present superiority and future perfectibility ("the science of culture," according to Tylor, "is essentially a reformer's science")—rhetoric that accorded nicely with U.S. schemes of social Darwinism and racial "fitness" as well as with long-standing national utopianisms.[17] Evolution, made to order.

Adams, however, while citing evolution as crucial to his own materialist historiography and aware of its aesthetic appeal (it was, he claimed, "the very best substitute for religion"), unconditionally refuses the optimistic terms of its popular representations (*Education,* 225). Rejecting any belief in a "doctrine of development"—whether biologically or socially engineered—he offers instead an alternative, pessimistic account of humanity's past, present, and future.[18] Perhaps most important for Adams is his conviction that evolution could offer only "pure inference," not evidence, of progressive "selection": "All he could prove was change," not improvement (*Education,* 230, 228, 230). Indeed, if evolution has a direction, *The Education* suggests, it is downward.

Echoing the "ebb[ing]" of "the life of the old earth" in the bleak furthest vision of Wells's *The Time Machine* (1895), this fear receives its fullest explication in *A Letter,* which focuses on an emergent countertradition in early-twentieth-century science that claimed that life did not have its own laws essentially different from those of inorganic matter—as vitalism presumed—but instead "obey[ed] the law[s] of thermal energy," like everything else (*Letter,* 150).[19] Founded on the observation that life functions according to energy gradients that equilibrate at death, this contention entails, for Adams, applying not evolutionary law but "the law of Entropy . . . to all vital processes" (*Letter,* 154). As Jennifer Fleissner has recently shown, Adams departs from the form of degradationism that, like its evolutionist inverse, posits the attractive unity and regularity of "uniformitarian" (*Education,* 399) descent in favor of a more chaotic decline ("change[s] of form broken by freaks of force . . . sheer violence, cosmic, chemical, solar, supersensual, electrolytic—who knew what?" [*Education,* 401]), but the decline is clear.[20] This is no small step: rather than evolutionism's promise of perfectibility, "the law of Entropy imposes a servitude on all energies," including the one most essentially human—"the mental" (*Letter,* 251).[21]

Although Stengers and others have shown this equation of entropy with degradation to be unnecessary (as evidenced by Clausius's early alternative account), Adams and many of his contemporaries believed entropy to entail the inevitable increase of disorder over time, making "man," presumably the latest addition to the world, "the most advanced type of physical decadence, no longer at the top but at the bottom of the ladder" (*Letter,* 166).[22] And *A Letter* presents, almost twenty years after Max Nordau's *Degeneration* and a decade before Oswald Spengler's *The Decline of the West,* such evidence of human "decrepitude" as "falling . . . birth rate[s] . . . multiplication of suicides . . . increase of insanity . . . 'habits' of alcoholism and drugs . . . and so on, without end" (*Letter,* 186–87).[23] The significance of these symptoms is far from abstract: *The Education* warns of an impending social "catastroph[e]" (499); "The Rule" declares that humanity's "movement [is] already retrograde" (303); and *A Letter* echoes claims that humanity is "destined to be rapidly extinguished" as it "face[s] accelerated extinction" (177, 166). Rather than fostering humanity's self-ennobling illusions, Adams offers evidence that "society [is] an organism in the only respect which seriously concern[s] historians:—It would die!" (*Letter,* 150). Little wonder that *The Tendency* contrasts the "pessimism" and "despair" (131) of Adams's "science of history" with "the cheerful optimism" of his contemporaries (129, 130).

First denying humanity's priority in a divinely ordered creation, then refusing the compensatory solace of evolution's promise of natural perfectibility, Adams also declines to make historiography into an instrument of sociopolitical intervention, rejecting the notion that anything can be done to avoid humanity's impending fate. Insanity may be rising and birthrates falling, but Adams offers no solutions; unlike Comte, Spencer, Fiske, and Marx, he does not promise a millennial future to ameliorate present suffering.[24] He can't, and for the same reason he believes his scientific historiography to be possible: the governance of history by invariable universal laws. *The Education* thus dispels "the fiction that society educated itself, or aimed at a conscious purpose" (483), noting that developments in human history that seemed "voluntary" were "in fact . . . as mechanical as the fall of a feather," with "all mankind . . . dragged on by an attractive power . . . as the planets obeyed gravity, or the trees obeyed heat and light" (484, 486).[25] For *The Education,* automatism gives the lie to autonomy, as all events become inevitabilities rather than achievements of individual or collective bodies.

Adams repeatedly acknowledges that such determinism, in which humanity is merely the mechanical reflex of its physical environment, could only have debilitating effects for a society formed in the belief that it is "child and heir to the infinite" (*Letter,* 231). "Above all," Adams writes of his dynamic theory, "it was profoundly unmoral, and tended to discourage effort" (*Education,* 501), because, as he puts it in *A Letter,* "human beings have always rebelled" "against [the] necessity" of "fall[ing] within the domain of physics. . . . For thousands of years they have stood apart, superior to physical laws" (229). Because his pessimistic cosmology brings this belief to a "crisis" (*Tendency,* 133), it must meet with derision from those entities—secular and religious—intent on preserving the sovereignty of the human self: "arranged in battalions . . . [were] the energies of government, of society, of democracy, of socialism . . . all striving to illustrate not the Descent but the Ascent of Man" (*Letter,* 157). "Society," Adams concludes, "cannot safely permit itself to be condemned to a lingering death" (*Letter,* 190), and so it cannot countenance a teaching premised on the "progressive enfeeblement of the will" (*Letter,* 207). If it were otherwise, if his theories were accepted, it would "mean the stoppage of teaching altogether" (*Letter,* 237) and "anarchy in art, in thought, and in society"; they would, in short, cause society to "commit suicide" (*Tendency,* 131).[26]

Given Adams's canny anticipation of the reception his theories were likely to meet, it should not surprise that they have never fared well in scholarly criticism. In the same measure that the "biographical" sections of *The Education* have been almost universally applauded, the dynamic theory and the *Degradation* essays have been consistently rejected. Most commonly, they have been either ignored or dismissed as a regrettable footnote to an otherwise compelling oeuvre.[27] Of those who do discuss Adams's scientific theories, many portray them as part of a vast "joke" on the universalizing pretensions of modern science.[28] Others understand them to be serious but primarily aesthetic, harnessing the power of prevailing symbols to create a modern mythology.[29] Even those who believe that Adams's search for a science of history was in earnest, and not simply a poetic vehicle, refuse to accept it. Many, for instance, psychologize it, ascribing its "queer malice" to some traumatic personal circumstance: dashed political ambitions; the commercial failure of his *History of the United States*; the erosion of his family's status; his wife Marian's suicide.[30] William Dusinberre goes so far as to conclude that Adams's late writings "verge on madness."[31] Similarly, Ronald Martin writes that Adams's dynamic theory represented "a real danger to sanity," and he ultimately denounces all of Adams's scientific theories as a "doom-metaphysics" resulting from "senile rage."[32]

What, we may ask, necessitates these dismissive responses? Why do parody, allegory, and pathology appear to be the primary options for interpreting Adams's science of history?[33] The answer perhaps lies in what these disparate readings share in common: an insistence on Adams's person as the proper hermeneutic lens for evaluating his theories. Though not without merit, such readings involve disregarding *The Education*'s and the scientific essays' rejections of exactly this prioritization of the authorial subject. To ascribe Adams's science either to his artistic will or to his subconscious is, in *The Education*'s schema, to mistake cause and effect. More fundamentally, this drive in the criticism to make Adams responsible for, rather than to, his deterministic theories signals a mostly unacknowledged investment in human exceptionalism and individual agency. When Adams's scientific writings are attributed to either parody or symbolism, they are viewed as expressions of his creative free will and thus as antidotes to the supposedly dehumanizing scope of modern science. Note Howard Munford's statement that Adams tried "to deflect attraction from the brutalities of science to a direction that would be *humanly* satisfying," or consider R. P. Blackmur's claim that Adams "had always the compulsion to put a *human* shape . . . upon . . . forces."[34] When, conversely, Adams's determinism is deemed to be genuine, humanity's status is preserved by making Adams into something less than normatively human: deranged, blinded by rage, or senile, but by no means competent to articulate a coherent cosmology.

Perhaps the most unabashed case for rejecting Adams's "inhuman" theories, however, comes from someone who actually takes them seriously. In his 1919 review of *The Education,* Robert Shafer criticizes what he understands to be the human costs of Adams's "suicidal doctrines":

> Everything recognisable as distinctively human is swept away, swallowed up in the anarchy of mechanical energies into whose presence the modern scientist proudly ushers us . . . a waste place inhuman and desolate beyond words to cry our woe. . . . Let us by all means admit that the universe is real . . . but let us not therefore deny our own humanity, distorting ourselves into mere helpless mechanisms. . . .
>
> Every man is aware of a different world within himself which is his sole possession, by virtue of which he is an individual. . . . And only the man who is conscious that there is a portion of his being which thus differs from, and even opposes itself to, his mortal constitution and its surrounding world of nature and society—only that man has become in the full sense of the word human.[35]

Defining the human as that which "differs from, and even opposes itself to . . . its surrounding world," then picturing the eschatological threat that the "suicidal" science of "modern[ity]" will visit on the earth, Shafer captures the implications of both Adams's position and much of the critical humanist resistance to it. Adams critics may differ in every other respect, but they share a collective unwillingness to countenance the danger to human personhood that Adams's texts represent. Like the evolutionists whom Adams indicts as incapable of renouncing the title of "lord of creation" (*Letter,* 231), they insist on saving Adams from his theories—and thus on exempting humanity and individuality from universal law.

The Education, however, rebuts exactly this state of exception. The essential distinction that critics posit between Adams's "scientific" historiography and his "subjective" autobiography is false: they are two facets of the same project, dual explorations of the fact that persons and cultures alike are products of processes inimical to their continued existence. *The Education* merely puts a human face on the *Degradation* essays, demonstrating universal principles through their action on a single person. Pressuring the Emersonian notion of representative men (and the larger American ethos of individualistic self-determination) to the point of implosion, Adams's text makes an example of Adams, picturing through his compelled life the impersonalization of our own.[36] *The Education* thus turns out to be a biography of force by proxy of a self, requiring us to make a distinction between auto-biography and autobiography, between a text generated by an "automaton" and one written by a person.[37] It is to an explication of this inhuman self-impoverishment that I now turn.

ADAMS ONCE WROTE, "No one shall persuade me that I am not a phase," and *The Education* bears him out.[38] As in Adams's scientific essays, *The Education* links universe and self through their shared determinism—and, as in the essays, the congruence results not in the self's transcendence but in its diminishment, material and spiritual. And one of the primary ways the text marks this deemphasis is through its recurrent insistence on Adams's passivity in relation to what can only tenuously be called *his* life. Thus, regarding the "education" of *The Education*: "Adams . . . was absorbing knowledge. He would have put it better had he said that knowledge was absorbing him. He was passive" (*Education,* 93), a claim echoed later: "Education . . . had done its worst. . . . He went on, submissive" (*Education,* 313). Time and again, rather than being the hero of his story, or even its causative force, "Henry" is merely "a helpless victim" prostrated by "the omnipotence of circumstance." What Adams writes of Adams prior to the Civil War is true of his life throughout *The Education*: "He could only wait for he knew not what, to send him he knew not where" (*Education,* 109).[39]

The Education, furthermore, leaves no doubt that the "passive obscurity" (*Education,* 502) that characterizes Adams's life results not from personal circumstance but from the "great mechanical energies" (*Education,* 238) and universal "force[s] all about him." Even Adams's "discovery" of the dynamic theory is attributed to the deterministic processes it analyzes:

> After so many years of effort to find one's drift, the drift found the seeker, and slowly swept him forward and back, with a steady progress oceanwards. . . . The process is possible only for men who

> have exhausted auto-motion. Adams never knew why, knowing nothing of Faraday, he began to mimic Faraday's trick of seeing lines of force all about him, where he had always seen lines of will. (*Education,* 426)

"Auto-motion" having ceased and heading "oceanwards," Adams, without even "kn[owing] why," passively "drifts" into the realization that both he and his "theory of the universe" are as mechanical in origin as Faraday's electromagnetism (*Education,* 427). The transition from traditional, self-centered agency to extrinsic compulsion is seconded by the passage's rhetoric. Notice, for instance, the ready slippage from "*one's* drift" to "*the* drift," a depersonalization that dramatizes the inhuman logic of substituting "lines of force" for "lines of will." The immediate consequence of this displacement is that Adams is revealed to be nothing more than the puppet of "forces [which] grasped his wrists and flung him about," neither intending nor superintending the direction of his education (*Education,* 494). With such a complete "exhaust[ion] [of] auto-motion," it becomes difficult to identify what role, if any, Adams plays in his life.[40]

Even authorship of the record of this displacement is denied Adams. First pushed off the page, he soon finds himself removed from his position behind it as well, a fact illustrated by *The Education*'s comment on the general "act" of writing and, implicitly, on its own genesis:

> The pen works for itself, and acts like a hand, modelling the plastic material over and over again to the form that suits it best. The form is never arbitrary, but is a sort of growth like crystallization, as any artist knows too well; . . . often the . . . pen runs into side-paths and shapelessness, loses its relation, stops or is bogged. Then it has to return on its trail, and recover, if it can, its line of force. (*Education,* 389)

In another context, this description might be dismissed as an innocuous, even clichéd, depiction of artistic inspiration. It might even be compared with the spiritualists' and the surrealists' contemporaneous investments in automatic writing.[41] In *The Education,* however, a "pen work[ing] for itself" is attributable to neither spirits nor the unconscious (both anthropocentric explanations) but rather, like the book and author thereby composed, to impersonal universal processes. Human agency is eliminated altogether. Note the passage's mention of a "hand," which, rather than being connected to a person, points only toward an (amputated) simile. Likewise the generic, immaterial "artist," who is granted only the capacity to "know"—as if secondhand—the process of textual production, not to participate in it.

Indeed, any possibility of human preference introduced by the phrase "the form that suits it best" is immediately undercut by the fact that the work's nonarbitrariness is not intentional but, à la Poe, "crystalli[ne]" in "growth."[42] Responsibility for both the loss and the recovery of narrative "shape" rests not with a guiding artistic consciousness but with a "pen work[ing] for itself" along "line[s] of force." The pen is thus not an extension of Adams's person, an object onto which his agency is projected, but rather a finger of the universe itself; alienated from himself, his hand is an external prosthesis to which he can claim no natural possession, part of what Jean-Luc Nancy calls "a body that we neither have nor are." "If I write," Nancy continues, "this strange hand has already slipped into my writing hand."[43] Undermining the committed author-centrism and humanist individualism of the Anglo-American literary tradition, the passage trades inspiration (and its metaphysics of circuitous-yet-lineal agency) for autonomic (not automatic) writing, leaving Adams outside and looking in on his own life. And far from signaling the birth of the reader or of discourse, the death of the author here portends the collapse of text and cosmos alike.

The replacement of agential subjects (artists) with inhuman objects (pens) is synecdochic of a larger topos in *The Education,* one that persistently fails to honor the difference between Adams's self and other things. The logic of the refusal stems from the fact that the dynamic theory finds human beings to be composed of the same materials and controlled by the same forces as the rest of the universe. Like the conflation of nerves and telegraphs in nineteenth-century literature and science, Adams's person is presented as an impersonal, networked technology.[44] It is telling, therefore, when *The Education* labels Adams "a mere instrument of measure, a barometer, pedometer, radiometer" of human and stellar history, because in the universe of *The Education,* Adams is always more instrument than instrumental, registering environmental effects without fully comprehending or mastering them (*Education,* 456).[45] And though this may at first appear to be figurative language, *The Education* insists on its literalness: these associations are finally not comparisons at all but equations, assertions more of identity than similarity. Calling Adams an "instrument" evidences a genealogical or homological relationship rather than a symbolic one because, owing to their equally determined operation, modern humanity really is "the child of steam and the brother of the dynamo" (*Education,* 466). Adams's posthuman thing-self is thus inconsequential but not irrelevant to *The Education*: its very want of autonomy and auto-motion gauges the power of cosmic law over the now ontologically flattened human. It is, paradoxically, what Quentin Meillassoux might call a living "arche-fossil," a product of events "anterior to the emergence of thought and even of life . . . *anterior to every form of human relation to the world.*"[46] As a sum of inhuman forces, Adams's person is such a speculative reality, pointing beyond itself to a cosmos prior to thought—"vast epochs of unknown time and change" (*Letter,* 172).

This is a crucial point, perhaps the most important of all for *The Education.* Conflating Adams with "instrument[s] of measure" illustrates that "subjectivity" is not the antithesis of "objectivity" but its origin and method; as in the work of his friend William James, but to opposite effect, if the self is inextricably a part of the universal process it seeks to theorize, then turning inward represents a viable means of exploring the outer world. Rather than problematic variables requiring emendation, therefore, Adams's putative life and the events of human history are to him as equally admissible as gravity and electricity for evidencing the operation of cosmic laws.[47] Because Adams is an object, his biography must be objective. This is why *The Education,* apparently only the record of an individual life, can claim to be a scientific "Study of Twentieth-Century Multiplicity" (*Education,* xxvii).[48]

It bears repeating, though, that Adams's inward turn does not comprise a phenomenology; for Adams, the study of mind constitutes not an end in itself but a way of grasping a greater truth. Thus, although *The Education* briefly considers whether the mind can offer the unity and meaning that the universe lacks, the text ultimately seizes on the revelation of late-nineteenth-century psychology "that the ψυχή—soul or mind—was . . . not a unit":

> The new psychology... seemed convinced that it had actually split personality not only into dualism, but also into complex groups... whose physical action might be occult in the sense of strangeness to any known form of force. (*Education,* 433)

Evidencing the same disintegration from "unity [to] multiplicity" as the rest of the universe, the psyche's "split" means that the "subconscious chaos" within the self is of a piece with the universal chaos outside of it (*Education,* 498, 433). Rather than an inviolate exception or last bastion of existentialist meaning, Adams's mind—only another "occult" "substance" ("Rule," 283)—proves the rule of universal deterioration: "Sanity was unstable artifice.... One saw no unity ahead—nothing but a dissolving mind" (*Education,* 434). For *The Education,* then, to turn inward is not to escape what lies without but to experience it as one's (but not one's own) self.

As the final exemplification of the universe's obliteration of selves, *The Education* repeatedly enacts Adams's death. We have already seen how the text imagines Adams to be "exhaust[ed] [of] auto-motion," left to be "swept... oceanwards" (a common figure for death in *The Education* and Adams's other writings). Similarly, Curie's discovery of the "anarchical" energies of radiation leaves Adams's "neck broken," a "pass[ing] away" depicted as the fitting expression of humanity's state within a universe of powers greater than itself (*Education,* 382, 317). And the remainder of *The Education* finds Adams "see[ing] nothing to make a return to life worth while" but "plenty of reasons for staying dead," making his education parallel that of "Nature," which "had... a singular sympathy for death" (*Education,* 320, 501). It is a sympathy that Adams shares outside of *The Education* as well. In letter after letter, he claims, "I am really dead,—stone coffin cold,—and I cannot go on with the old life," or "I am dead to the world;—dead as Adam and Eve, only just not yet buried.... I hate to go about as though I were real. Pretending to be alive is a positive swindle.... Respect my phantasmodesty... and let me slink back to my place like a ghost, to find out in silence and peace whether I am still a little bit alive."[49] Similarly haunted by a specter of what should be his self, *The Education* is Adams's "last Will and Testament" to the diminution of human life within a chaotic *and* deterministic universe.[50]

Such a "necrospective" vision both accords with and departs from posthumanist theorizations of chaos and evolution.[51] To take one example: in revealing himself to be nothing more or less than an assemblage of various states of "matter-energy," Adams anticipates Manuel DeLanda's undermining of the false essentialization and "reified totalit[y]" of the individual.[52] Like Adams, DeLanda insists that persons are "emergent propert[ies]" of the organic and inorganic strata that compose them.[53] Moreover, both Adams and DeLanda extend this insight beyond the individual to implicate the entire "human species . . . as . . . a contingent historical production," the inevitable consequence of which is to "allow physics to infiltrate human history."[54]

Yet where the emphasis in DeLanda, building on Ilya Prigogine's repurposing of entropy, is on "matter's inherent creativity," on how assemblages come to be through nonlinear, nonteleological processes, Adams's focus is on decreation, on how orders—personal, social, cosmic—dissolve.[55] Perhaps these positions represent two sides of the same coin, reflecting what James would call only a difference of "temper."[56] It is, however, a difference that matters. Although DeLanda explicitly cautions that thermodynamic evolution neither can "be viewed as . . . [moving through] progressively more sophisticated stages . . . that culminat[e] in humanity as its crowning achievement" nor can "guarantee . . . a golden age ahead," he nevertheless locates "a new kind of hope" in the "positive, even joyful" project of "*destratify*[*ing*] *reality itself*"—a project based in finding "the pragmatic uses of" thinking in terms of heterogeneous processes.[57] In this, DeLanda is joined by Elizabeth Grosz, Keith Ansell Pearson, and numerous other theorists seeking to "reconceptualiz[e] the dynamics of political change and social and cultural upheaval" by basing them in "the unpredictable . . . [and] nondeterministic" events of temporal and material flux.[58] By converting universal indeterminacy into a praxis of personal and collective liberation, they open new horizons of "endless becoming."[59] It is an idea, of course, that originates not in our age but in Adams's, where Bergson, James, and others link the "creative evolution" of the cosmos—"the most intimate secre[t] of life"—to the revitalization of ourselves.[60]

Adams, however, refuses such comfort. Where others believe universal process to be inherently democratic, Adams sees only the despotism of nonhuman powers (recall the title of *The Degradation of the Democratic Principle*); where they make cosmic principle useable, he despairs of our incapacitation. For Adams, unity is not harmony and anarchy is not liberty because both are beyond our control. What's more, Adams claims that both must end in our death. Renouncing the allure of serviceable alterity, *The Education* is the tombstone for the notion that "the universe . . . [is] a sort of joint-stock society" underwriting our freedom and ourselves.[61]

But doesn't *The Education*'s existence belie its message? Doesn't the fact of its utterance trouble the fact that it utters (i.e., Adams's death)? If Adams is removed from the text, didn't he do the removing? Dying as character, doesn't he survive as author or narrator—existing, as it were, in the margins? Is it not the case, as de Man—echoing Derrida—argues in "The Rhetoric of Blindness," that the "assertion of [the] absence [of the self] reintroduces a form of selfhood" and that this implies "a duplicity within a self that willfully creates a confusion between literal and symbolic [death] in order to achieve self-transcendence as well as self-preservation"?[62]

Such questions seem especially pertinent given the frequency and nonfinality of *The Education*'s death scenes, scenes which, Poe-like, never quite commit themselves to burying Adams once and for all. Recall the moment that has Adams "see[ing] nothing to make a return to life worth while . . . [but] plenty of reasons for staying dead." What's bizarre about this formulation is its representation of a posthumous consciousness contemplating both its own (non)existence and—even as it rejects the option—the *possibility* of "a return to life." Likewise, the last paragraph of the book, which, after professing that "it was time" to die, then imagines "some day" in the future when Adams "might . . . return" (*Education,* 505). But the ambivalence is perhaps clearest in the letter that claims that Adams is "dead . . . just not yet buried" and then continues, "Pretending to be alive is a positive swindle. . . . Let me slink back to my place like a ghost, to find out . . . whether I am still a little bit alive." Because we are given contradictory information (Adams is dead, then he feigns life, then he's unsure), it's difficult to categorize exactly who or what is making these statements; but that *something* still exists is made obvious, paradoxically, by the very claims to the contrary.

This seeming survival of death—in conjunction with the paradox of an autobiography written in the third person—has led many readers of *The Education* to insist on Adams's presence behind the text.[63] And once Adams is reinstated as author of his life, it's not far to abandoning *The Education*'s determinism altogether—not far to claiming, as we have seen, that the whole of the book is an exemplary exercise in irony, a work that does not mean what it says or says one thing and does another. Here we can see a fundamental continuity between the modernist claims that make *The Education* a vast joke and poststructuralist accounts of the text. Howard Horwitz, for instance, echoing Blackmur, argues that "Adams abandons the search for truth" after discovering modern science's revelation that we can never have objective knowledge of the universe; instead, we can only "project" "wholly subjective" and self-consciously ironic orders to satisfy our desire for meaning—"grand narrative[s] of... reflexive self-confirmation that constitut[e] human identity."[64] According to Horwitz, Adams's theory of history is just such a narrative, an ironic performance whose fictional universality stands in for the cosmology it only apparently posits. Similarly, Gregory Jay contends that the dynamic theory manifests the relativistic "play" and "ironic self-fashioning" of a Nietzschean artist asserting "human truth[s]" rather than inhuman laws.[65] For both Horwitz and Jay, then, irony is ultimately *humanizing* because it frees us from the dominion of objective truth. Their assertion of the groundlessness of grand narratives—physical and metaphysical—sanctions Adams's own ironic articulation of a determinism that he is not determined by.

If this sounds similar to the optimism of the creative evolutionism considered earlier, it's because it is: both situate the human subject within impersonal, inhuman systems (whether material or discursive) to deconstruct its givenness and priority, but both then marshal these irruptive forces in the name of a free play that allows for the return of our agency, if not ourselves.[66] In both, furthermore, the systems themselves are the means of our liberation, where the escape is not from process per se but from the false orders and limitations we impose on it. To fix the game, we need only learn to abandon the fantasies of stasis and separation; if we accept the prison, we might rule it. In this, posthumanism and poststructuralism can surprisingly echo the anthropocentrism of Adams's humanist contemporaries (albeit in revised form): if we can no longer *find* the meaning of the human in the world, then we can *produce* it. Either way, we *use* evolution to create our true—because truly free—selves.

Rather than deny *The Education*'s irony, however, I want to propose a different understanding of its function within the text. Horwitz, Jay, and Adams would agree that irony results from humanity's inability to materialize the metaphysical orders it projects onto the universe, but where Horwitz and Jay contend that such failure results from the *absence* of universal law, Adams understands it as the effect of this law's *revelation*. Irony is for Adams not the privilege of an exempted self but the consciousness of the forces that erase it. The inevitable outcome of the encounter between humanity's central assumptions (self-determination, ontological significance, a benign universal order) and physical reality, irony is an awareness—itself compelled—of the unbridgeable gap separating the two. Thus, instead of providing an opportunity for presence, play, or performative self-fashioning, *The Education* makes irony the realization of the negatively free self's impossibility within a deterministic universe, meaning that Adams is not the agent but the object of irony. De Man would agree: irony can never "benefit the self" by being "a preliminary movement toward a recovered unity" because "the act of irony allows for no end. . . . It can know . . . inauthenticity but can never overcome it."[67] This is not to say that *The Education* is not, in its own way, self-centered (if only in the pessimistic affect of its adamant third personhood); it is only to suggest that the self at the center of *The Education*—unlike those eminently modern subjects at the heart of nineteenth-century novels—is hollowed of its ontological significance by the irony of the terms of its existence.[68] We can know the truth of ourselves in such a universe, but it will not set us free.

To illustrate, let us return once again to *The Education*'s death scenes, which, as we have seen, result in something more like evisceration than execution. Yet rather than regarding such nonfinality as an attempt on Adams's part to escape the law he theorizes, we might see it as the paradigmatic state of being in a deterministic universe. Recall that the Adams of these passages is "like a ghost," neither fully "dead" nor fully "alive" because at least partially ("a little bit") both, existing in a liminal space somewhere between being "buried" and being "real" (as de Man writes of Wordsworth's *Essays upon Epitaphs,* Adams "moves, without compromise, from death *or* life to life *and* death").[69] "Blur[ring] the boundary between life and death," he is prosopopeia made flesh, a corpse who "speaks, but . . . only of the grave, not of resurrection."[70] Although an "Adams" still exists, then, it's just barely, like an echo of life rather than the genuine article: an entity in limbo, a trace, something precisely ghostlike because neither of this world nor beyond it—what Derrida calls "hauntological."[71] It is, in short, exactly the kind of spectral existence that defines Adams's existence throughout *The Education,* which means that his "deaths" are synecdochic of a life denied full presence. Made and unmade by nonhuman powers, Adams's is a half-life. If, therefore, he resembles a zombie or a machine—in motion but nonvolitional—it is because the universe has made him so. Determinism doesn't mean the complete absence of "Adams," only the death of what we would normally regard as his self.

Asking whether Adams is or is not "in" *The Education* thus turns out to be the wrong question because the universe places Adams on the line between presence and absence, in the interstices of being and nonbeing; or, as conceived by Søren Kierkegaard, in the quintessential space of irony. Writing that "the ironist . . . is lighter than the world," Kierkegaard claims that "irony as infinite and absolute negativity is the . . . weakest indication of subjectivity."[72] The ironist, both of and not of the world, is an apparition registered only by "the dead silence in which irony . . . haunts" (*Concept of Irony,* 258). This "qualification of subjectivity" (*Concept of Irony,* 262) results from irony's inevitable ironization even of the ironist, as it

> is a nothing that devours everything, and a something one can never grab a hold of, something that is and is not at the same time. . . . Just as irony surmounts everything by seeing its misrelation to the idea, so it capitulates to itself, because it continually goes over and beyond itself. (*Concept of Irony,* 131)

For Kierkegaard, irony is not a positive practice of the self so much as a "devour[ing]" of oneself in an infinitely reflexive regress.[73] And *The Education* is the disorienting product of exactly such a "misrelation," a text in which Adams's self is forced by its encounter with the universe to "continually go over and beyond itself" but never to arrive at anything else.

Ultimately, however, *The Education*'s irony is even more radical than Kierkegaard's. For Kierkegaard, ironic detachment originates in the ironist's subjective distance from the world, and it serves the function of advancing world history by loosening the current age's adherence to the status quo. For Adams, irony is not the result of *detachment* but of a radical *integration* that works to erode not only the ironist but also history as such. Although Adams would agree, then, with de Man's claim that autobiography is a signal instance of the defacing of all identities, he would insist that such an irony results from the impersonality not only of language but also of *matter*, of our self-alienatingly material composition. According to Adams, "we" are literally deconstructed by the non-self-identity of our *physical* persons, by their untenable continuity with deforming cosmic powers. And if this is a referential logic—if the textual attenuations of Adams echo the universe's—then it is one that paradoxically relies for its success on the ways in which the damning silence of the universe finally becomes our own, leaving us with "the dead silence in which irony . . . haunts."[74] We are left, that is, with the silence of a life indistinguishable from its absence, the echo of a death that outlives—by preceding—its birth.

In this context, it is significant that *The Education* both names a chapter "Silence" and presents silence as the best way "to express what is incapable of expression" (*Education,* 504), ultimately concluding that "dispute was idle, discussion was futile, and silence . . . the mark of sense" (*Education,* 501). Having learned this lesson, *The Education* itself falls silent soon thereafter. Elsewhere, however, Adams continues to speak silence, even intoning, "I believe God to be Silence."[75] And when we note that Adams writes admiringly in numerous places of the Buddha's silence, we can understand the full meaning of his private name—"Buddha"—for the memorial sculpture he commissioned Augustus Saint-Gaudens to make for Marian's grave.[76] It was only a private name, though, because against such popularly attributed labels as "Grief," Adams insisted on the sculpture's "universality and anonymity," and he mandated that no names or dates be inscribed on it. As a figure, in Adams's words (recounted by his friend John La Farge), of "the acceptance . . . of the inevitable," the sculpture—shrouded, impersonal, eyes closed—is a monument to Adams's belief that we are "a crushed humanity": ephemerally present as things, forever absent as selves.[77] (Figure 3) Given such a belief, it should not be surprising that Adams expressed admiration for Michelangelo's Slaves or that he directed Saint-Gaudens to study Michelangelo for inspiration.[78]

It is here that the challenge of Adams's science announces itself most insistently. If, despite their failed prophesies and discredited formulas, *The Education* and *The Degradation* usefully intervene against the naive and potentially pernicious elements of anthropocentric evolutionism, they do so in a manner that necessarily disallows their mobilization for constructive posthumanist alternatives. In dismantling the notion that humanity is ontologically removed from or superior to the world, Adams also forecloses the possibility that we may benefit from our union with it. He offers no liberation, no escape, no hope that we might restore our lordship. "Profoundly unmoral . . . and tend[ing] to discourage effort" (*Education*, 501), Adams's theories proclaim only their own inconsequence within a framework of imminent extinction: "The triumph of this teaching is the ultimate degradation of . . . the teacher as well as of the pupil and the universe" (*Letter*, 208).

Of course, the teaching's power inheres precisely in this negativity, in its unyielding refusal of our attempts to humanize the cosmos. From the perspective of processes inhuman in their spatiotemporal scope and effect, processes that all but ensure our eventual extinction, such attempts are inconsequential at best and violently destructive at worst. As Adams was keenly aware, the legacy of our trying to impose self-interested orders onto the world is one of "physical violence": "history shows man as alternately insane with his own pride of intellect, and shuddering with horror at its bloody consequences" (*Letter*, 206–7). Here the supposed evidence of our natural superiority and perpetual progress—the mechanization of war, "the destruction of life . . . on a scale that rivals [the] operations of nature," the "deforest[ation of] the planet," the stripping of "coal-oil and gas out of the earth" (*Letter*, 217, 216, 216)—is converted into testimony that "man is a bottomless sink of waste unparalleled in the cosmos, and can already see the end of the immense economies which his mother Nature stored for his support" (*Letter*, 218).

FIGURE 3. Augustus Saint-Gaudens's *Adams Memorial* (1886–91). Rock Creek Cemetery, Washington, D.C. Courtesy Jerry L. Thompson/Art Resource, New York.

Yet, and this is the self-consistent limit of Adams's determinism, even if man "startles and shocks . . . himself, in his rational moments, by [the] extravagance" of his devastations (*Letter*, 217), he is helpless to stop them because, "from the physicist's point of view, Man, as a . . . natural force, . . . [has] no function except that of dissipating or degrading energy" (*Letter*, 216). We have an "*instinct* for destruction"

(*Letter*, 217; my emphasis), which means that however much we may lament it, we not only can't reverse entropy, we are its essence: "As an energy [man] has but one dominant function:—that of accelerating the second law of thermodynamics. So far as his reason acts as an energy at all, it is a miraculous invention for this purpose" (*Letter*, 230). Our destruction of the earth isn't unnatural or immoral, therefore, it's written in the stars—as is our Lucretian fate: with all of "the higher animals," we, too, will soon "pass, like dreams" (*Letter*, 219).[79]

In this peering into the abyss of nonhuman time (far deeper than the human bounds of Wai Chee Dimock's "deep time"), Adams anticipates recent attempts by Ray Brassier, Mark McGurl, Claire Colebrook, and others to think the implications of extinction on our material, social, and philosophical orders.[80] Refusing to humanize the threat that such extinction poses by framing it within the context of our ability to cause or prevent it, these authors join Adams in dashing anthropocentricism on the rocks of our own obsolescence. Thus undone, we may now "conside[r] a world that is not ourselves and a force that cannot be returned to the human."[81] Of course, the cost of this relativization—and here Adams is darker even than Poe—is not only literal extinction but also the death of all naturalized ethics, as we can no longer "assume an ethos, a proper way of being, community, 'we' or humanity that would be [our] ground and value."[82] We may, for instance, turn Adams's theories against his own racism and anti-Semitism, but, if we remain true to his vision, we can do so only by universalizing them, by claiming that we are all degraded; his theories may deauthorize colonization, but they cannot justify decolonization.[83] They are quietist, but by ontology rather than, as Paul Bové eloquently has argued, by existential choice.[84] If Adams's universe offers anything to us, then, it must be found within this radical passivity, this inescapable quiescence, this overwhelming silence that Adams insists is our fate within a universe not our own.[85]

Cataclysm may be the inexorable outcome of Adams's determinism, but it is possible to imagine an antianthropocentricism that denaturalizes our violent attempts at mastery without ending in literal extinction. I thus turn in the remaining chapters to an African diasporic worldview that, though not reducible to its encounter with New World slavery, offers a powerful challenge to the conjoined metaphysics of racism and anthropocentrism. Destabilizing the notion of the masterful human subject but without recourse to a final, damning apocalypse, the hoodoo cosmology avoids Poe's and Adams's godlike hubris of tying the universe's fate to our own. Less fatalistic because less teleological, it allows for the possibility of a livable—because uncertain—future. Granted, it may not be "our" future, but our absence would be due to transformation rather than annihilation and thus would not preclude a new community of "we."

3

"BY AN ACT OF SELF-CREATION"

On Becoming Human in America

THE PREVIOUS TWO CHAPTERS introduced cosmologies that countered the anthropocentric optimism of mesmerism and popular evolutionism by making the human subject's fusion with cosmic processes—Poe's gravity, Adams's entropy—fatal rather than elevating. Yet Poe's and Adams's contestations of human ascendancy were, from the perspective of actual, lived experience, largely academic; as white men, they were not in danger of being denied their humanity. Their auto-alienations were not socially mandated, their antagonism toward humanism not the result of being excluded from it. Indeed, the totalizing nature of their theories could be seen as formally analogous to, even predicated on, the assumed universality of white male subjectivity in nineteenth- and early-twentieth-century America.

The next three chapters, however, consider the literal dehumanizations of U.S. slavery and its racist afterlives in the context of an African diasporic spiritual tradition that explodes their metaphysical justifications. Like the cosmologies of Poe and Adams, this tradition (embodied in the voodoo, hoodoo, rootwork, and conjure of the black American South) troubles absolute human–nonhuman and self–other divides—and thus white–black ones as well. The result is the disarticulation of white ontological privilege, but not through its extension to those previously denied it (as in many slave narratives); instead, the privilege—and the violently exclusionary category of the master self that it subtends—is abandoned altogether. Equality still exists, then, but it is based in abjuring rather than expanding liberal subjectivity because here no one is hierarchically human, no one self-possessed. Correlatively, no thing may be reduced to the death sentence of "bare life" or mere, passive nature.[1] Nothing is ontologically killable because everything is, at least potentially, a living relation—not because everything is related, per se, but because everything *is* a relation. "I," "we," you," "it": relations all, all the way down.

We can see this when we look at the broad outlines of the diaspora cosmology itself, which can be traced, at least in part, to elements of African traditional religions (ATRs) that survived the Middle Passage.[2] It is, of course, impossible to collapse the two traditions, not least because of the rich diversity—and dynamic, syncretic creolization—of spiritualities on either side of the Atlantic.[3] I am not interested, however, in identifying a single, homogeneous, wholly other Afrocentric belief system to pose against an equally monolithic Western one. To do so would risk, among other things, perpetuating what Curtis J. Evans has identified as the pernicious "romantic racialist conception" of "black innate religiosity."[4] Rather, I follow the seminal scholarship of Albert J. Raboteau, Lawrence W. Levine, Robert Farris Thompson, and others in noting the resonance between *specific* beliefs found in both West Africa and black North America, beliefs that include a recognition of life and consciousness in supposedly inanimate places and things;[5] a consequently attenuated category of the human;[6] a displacement of individual identity, with multiple souls inhabiting a "single" person;[7] possession by spirits;[8] "porosity" of the physical body to infestation by lizards, snakes, and insects;[9] and manipulation or death by dolls, conjure bags, and metamorphosis spells.[10] I will use the shorthand of a conjure, voodoo, or hoodoo cosmology to describe this constellation of beliefs, but I by no means intend it as a comprehensive account of diasporic religion in general or even of conjure, voodoo, or hoodoo in particular (though the latter three largely overlap via what Christopher Fennell labels "ethnogenic bricolage," each has its own distinct foci).[11]

Considered on their own, however, the beliefs are striking. Blurring the boundaries between inside and out, person and environment, self and other, they indicate not—as one might expect from standard critiques of Enlightenment being—an ecstatic, liberating union but rather a radical, often painful decomposing in which "we" really do lose "our" selves. And it is a dispossession no more escapable than wholly benign because it is simultaneously the condition of our existence and the principle of its estrangement. From this perspective, the conjure cosmology's erosion of man-made/man-making ontological difference works to expose the aporias of a metaphysics whose fundamental orientation toward the world is premised on either a self-serving distantiation or a self-serving subsumption.

Yet unlike Poe's and Adams's universes, the hoodoo cosmos does not exact actual extinction as the price of the death of Man. Universal forces here divest the human self of its imperiousness through *ongoing and open-ended* processes rather than a final—and finally anthropocentric—apocalypse. People still die in this cosmos, of course, but they also survive, meaning that the real threat is to our sovereign autonomy, our impossibly independent-yet-central existence. Even this overthrow is more of a by-product than a design because there is no telos, positive or negative, no mechanically monocausal law, only a dense web of agencies working in complex relations of resonance and interference. The emphasis is on change, metamorphosis, and transformation, not the essentially static working out of a predetermined end. Put another way, the cosmology is nonfutural, which is to say that it allows for the possibility of the future precisely because we cannot fully master or know it. Unmoored from the free will–determinism dichotomy, there is a chanciness to this universe, as intention and accident, order and chaos, blur almost to indistinction. Universal forces still obtain, but they are contingent and *locally* manifested.

It is, in short, a cosmology that anticipates much recent ecocriticism and actor–network theory (ANT) in theorizing co-constitutive yet nonharmonious imbrications of persons, places, and things. ANT's emphasis on approaching the human and the nonhuman symmetrically is particularly relevant, as both it and the conjure universe compel our recognition that action—even what we believe to be *our* action—"is overtaken or . . . *other-taken*," meaning that "we are not alone in the world."[12] However, rather than attempting, impossibly, to apply ANT or some other discourse to the hoodoo cosmology (thereby making the latter subordinate, secondary, or merely exemplary), in what follows, I focus on the lived experiences, narrative representations, and immanent philosophical implications of the cosmology itself, while also addressing how these dimensions resonate, or not, with current scholarly concerns.[13] As we shall see, this tradition offers its own poignant presentation of the enmeshment of human and world as "a dynamically developing cluster of forces" rather than a "static and unchanging being."[14] In so doing, it articulates a vision of mutual indebtedness and common, heterogeneous flesh that refutes both isolationism and holism, even as it suggests that some forms of relation might be less harmful than others; unmappable onto the current ethicopolitical landscape but not utopian, incapable of naturalizing itself beyond its own terms but not without grounding, it is an alternate universe inhabitable only by an alternate, perhaps impossible, "us." It finally can't compel "us" to change, it can't even tell us how or why we would, but it nonetheless insists that "we" have always been other than we thought.

More of an overture than a self-contained argument, the present chapter begins the exploration of this cosmology with a study of its complicated rejection by Olaudah Equiano and Frederick Douglass, before transitioning to its equally complex reception in the post-Reconstruction United States, thereby setting the stage for my discussions of Charles Chesnutt and Zora Neale Hurston in chapters 4 and 5.

LOCATING AN AFRICANIST CRITIQUE of certain Euro-American metaphysical traditions is, of course, a well-established practice, with particular attention having been devoted to the ways in which African American slave narratives deconstruct the philosophical justifications of slavery. In *The Black Atlantic,* for instance, Paul Gilroy recruits Frederick Douglass's autobiographies to the cause of resisting the process by which "European particularisms are . . . translated into absolute, universal standards for human achievement, norms, and aspirations." First simply presenting Douglass as aware of "the incomplete nature of the Enlightenment project," Gilroy goes on to suggest that Douglass's critique is even more radical.[15] Included as a part of what Gilroy calls the slave's questioning of "the Enlightenment . . . idea[s] of universality, the fixity of meaning, the coherence of the subject, and, of course, the foundational ethnocentrism in which these have all tended to be anchored," Douglass is the primary example of a "slave perspective" that "necessitate[s] a critical stance on the discourse of bourgeois humanism" and its role "in the rise and consolidation of scientific racism." It is crucial for Gilroy, then, that Douglass not be seen as merely agitating for the greater inclusiveness of Enlightenment ideals because, were this the case, it would negate Douglass's availability for illustrating the fact that European "humanism cannot simply be repaired by introducing . . . black folks who had previously been confined to the intermediate category between animal and human."[16]

To support his claim, Gilroy turns at length to the scenes in *My Bondage and My Freedom* surrounding Douglass's conclusive fight with his brutal overseer, Covey. When, for instance, Douglass claims that after "repelling the unjust and cruel aggressions of [the] tyrant," he "had reached the point . . . at which [he] was *not afraid to die,*" Gilroy argues that the contention represents an "alternative to Hegel": in preferring death to bondage, Douglass subverts the logic whereby slaves are defined by their acquiescence to their masters.[17] More specifically, Gilroy believes that Douglass's reversal outlines a space of absolute "freedom" directly inspired by the "Africentric . . . alternative" of "black spirituality," which "articulates a principle of negativity that is opposed to the formal logic and rational calculation characteristic of modern western thinking."[18]

Gilroy's evidence for Douglass's Afrocentrism comes in part from the latter's encounter with Sandy, whom Douglass describes as "a genuine African . . . [who] had inherited some of the so called magical powers, said to be possessed by African and eastern nations."[19] Situating his analysis in a longer commentary on Douglass's interest in ancient Egypt, Gilroy underscores the fact that Douglass "tr[ied] the African" "alternative" to his "Christian scepticism" and "book learning" by taking "the root" Sandy offered him for protection against Covey; and he is careful to note that Douglass admits to "'a slight gleam or shadow of . . . [Sandy's] superstition . . . fall[ing] upon'" him as he does so. For Gilroy, this illustrates an acceptance of "a pre-rational, spiritual mode of African thought"—and thus an instance of an African American author "dr[awing] deliberately and self-consciously on premodern . . . symbols that gain an extra power in proportion to the brute facts of modern slavery."[20]

What Gilroy crucially omits, however, is that in Douglass's estimation, "all this talk about the root, was . . . very absurd and ridiculous, if not positively sinful. . . . I had a positive aversion to all pretenders to '*divination.*' It was beneath one of my intelligence to countenance such dealings with the devil, as this power implied." Moreover, though

Douglass confesses to being momentarily compelled by the "earnest[ness]" of Sandy's "superstition," the root soon proves worthless when it in fact fails to prevent Covey from attacking him. Consequently, Douglass is quick to discard the root and, in classic American fashion, defend himself: "I now forgot my *roots,* and remembered my pledge to *stand up in my own defense. . . . I was resolved to fight.*"[21] Self-reliant force of will, not "superstition," African or otherwise, secures Douglass's victory.

Thus, despite Gilroy's argument to the contrary, which I have engaged at length because of the power and influence of its insights, I would contend that Douglass's overcoming of the fear of death in the pursuit of freedom has a more proximate source than African spirituality. Is it not likely that his claim of a "manly independence" begotten because "[he] was *not afraid to die*" strategically recalls Patrick Henry's "give me liberty or give me death!"? And what else is Douglass's reference to a "tyrant" but a simultaneously castigating and beseeching arrogation of Thomas Paine's *The Rights of Man*? If masters and slaves are on Douglass's mind in 1855, might not citizens and despots be as well—as was the rhetoric of slavery on the minds of the colonial patriots? This is not to say that Douglass's argument for equality wasn't revolutionary; it's only to note that it was also decidedly Revolutionary.[22] Douglass strategically plays America's founding discourses in ironic, accusatory counterpoint to the realities of contemporary slavery. And it bears emphasizing that the irony works not against the discourses themselves but against their "shameless hypocrisy"—against the incomplete realization of their promised universality.[23] Rather than an "Africentric . . . alternative," Douglass regards the nation, *his* nation, as responsible for both his bondage and his freedom.

By invoking a Euro-American rhetoric of freedom, Douglass necessarily claimed the status of human selfhood to which it exclusively applied. Such status was the very thing at issue, of course, because centuries of religious, political, and economic justifications of slavery had made the nonwhite slave something less than fully human—indeed, the antithetical other against which humanity was defined. As Joan Dayan notes,

> the rules for controlling slaves in the colonies depended on the enlightenment strategy by which humans ruled the universe of

> things, including, through a fantasy of reification, slaves. Descartes's experiment with himself to establish the idea of the white universal subject is thus recovered as a collective experiment to legally produce black nonpersons in the New World.[24]

To claim America's independence as his own, therefore, Douglass needed to prove nothing less than his race's humanity, its distinction from the category of thingness to which it had been brutally assigned.

The enormity of the gambit is difficult to overstate, but Douglass was by no means alone in undertaking it. Numerous scholars have argued that slave autobiographies as a whole can be defined by their defiant articulation of a humanity forcefully refused them.[25] Indeed, for Henry Louis Gates Jr., authorship is itself the most poignant demonstration of the slave autobiographers' personhood:

> The texts of the slave could only be read as . . . the slave's . . . reversal of the master's attempt to transform a human being into a commodity. . . . The slave wrote . . . to demonstrate his or her own membership in the human community.
>
> . . . Slaves and ex-slaves met the challenge of the Enlightenment to their humanity by literally writing themselves into being through carefully crafted representations in language of the black self.

Through the demonstration of literacy ("the sole sign of difference that separated chattel property from human being"), Gates summarizes, "a black person could become a human being by an act of self-creation."[26] If in the West to write is to be, then for an ex-slave to write himself as a human being and not a commodity, as a "Thou" rather than an "It," is to countermand his relegation to the nonexistence dictated by racist philosophy and practice.[27]

"I WAS A MAN NOW": A LIFE WORTH KILLING FOR

Though Srinivas Aravamudan, Valerie Smith, and others have cautioned that literacy did not, in fact, guarantee the recognition of African Americans' humanity in slave society—and that the idea itself risks "reiterating the evolutionary narrative so dear to early abolitionists, that of humanizing the African" according to the standard of "the Eurocentric self"—Douglass would have agreed with Gates.[28] He insisted, for instance, that he author both his life and its telling to better demonstrate his (and his race's) equal humanity: "There are special reasons why I should write my own biography, in preference to employing another to do it. . . . It is alleged, that [enslaved people] are, naturally, inferior; that they are so *low* in the scale of humanity, and so utterly stupid, that they are unconscious of their wrongs, and do not apprehend their rights. Looking, then, at your request . . . I part with doubts and hesitation, and proceed to furnish you the desired manuscript."[29] By announcing in print his consciousness of both his "wrongs" and his "rights," Douglass defies the assumptions of even those Northern white abolitionists who saw him as only an unthinking "'*chattel*'—a '*thing*' . . . [an] *it* [that] could speak."[30] Self-authorship elevates him onto the same "scale of humanity" as them, that of rational, conscious, ethical subjectivity. In this regard, Douglass's argument for the equality of African Americans is confirmed not only by its content but also by the mere fact of its existence. It is performative, actualized by its utterance.

Importantly, however, Douglass does not allow his writing to have the last word in his humanity's defense. Rather, he stakes his ultimate claim to personhood on his conclusive fight with Covey. If by being beaten by Covey, Douglass is made to exchange his "manhood for the brutehood of an ox," then only by giving a beating back to Covey is Douglass restored as a man: the "battle," as Douglass describes it, "was the turning point in my 'life as a slave.' . . . I was a changed being after that fight. I was *nothing* before; I WAS A MAN NOW."[31] From the nothing, or "*no-thing*," as James Cone labels it, of slavery to the humanity of "manly independence," Douglass is literally "a changed being" after besting Covey—indeed, it is in this way that he is first recognized as a "being" at all.[32] No longer a thing to be determined by others, he is now that most paradoxical yet familiar of nineteenth-century entities: a self-made man.[33] *My Bondage*, and the fight with Covey in particular, is thus the story of one man's *becoming* (belatedly) human, of the transition from being another's property to owning—in all senses of the word—one's self (*my* bondage, *my* freedom).

Contra Gilroy, therefore, Douglass believes that "the Enlightenment project" *can* "be repaired by introducing ... black folks"; rather than "critiqu[ing] ... Enlightenment universals [from] outside the philosophical repertoire of the West," Douglass critiques only these universals' untenable exclusions of persons of African ancestry.[34] Far from questioning "the coherence of the subject," Douglass's autobiographies forcefully demand that he be recognized as a self equal to any other (note the emphatic textual formatting of "I WAS A MAN NOW"). Self-determination, patriotism, Christian ethics, manly independence: Douglass refers to each with approbation—and carefully underscores the fact that they appear more frequently in his own life than in the majority of the whites he encounters. This is only to acknowledge the accomplishment of what *My Bondage* and many other slave autobiographies set out to do. It is not to deny the devastating force of their unanswerable critiques of slavery's inhumanity, nor is it to discount the practical necessity or political expediency of articulating these critiques in the manner that they did (being accepted as human was, after all, a life-or-death matter). It's simply to say that theirs was never intended to be a total "rebellion" against Enlightenment tenets of self and world or an absolute "indict[ment] ... [of] the [Western] metaphysical system," as some have claimed.[35] Douglass thus broadens but does not abandon the category of masterful personhood; the standard is met, not abolished, the line displaced, not erased. This in itself poses a sharp reprimand—it disproves the racist assumption of the incapacity of African Americans for such an enterprise—but it does so by repurposing rather than discarding the mechanisms of their subjugation.

Given the alternative Douglass faced, it would be preposterous for anyone (especially a twenty-first-century white male academic) to argue that he should have done otherwise. In the absence of such toothless judgments, however, we might nevertheless acknowledge that Douglass's success at proving himself equal to his masters' standard of humanity, necessary though it was, involved costs as well as gains. Even, that is, as we praise his powerful negation of his supposed inferiority, we can be cognizant of the problems associated with his "becoming a westernized metaphysical self."[36] This may seem an abstract concern relative to the dehumanizations of slavery until we realize that Douglass's achievement of personhood paradoxically threatens to naturalize the very logic that justified his enslavement. The danger arises not only because "constitutional personhood has . . . always been defined in proximity to slavery" (Barbara Johnson) and continues to be routed through "the domain of property" (Stephen Best), or even because "terror and . . . violence" inhere in the ex-slave's becoming a "self-possessed individual" such that "the recognition of humanity and individuality acted to tether, bind, and oppress" rather than liberate persons from "savage encroachments of power" (Saidiya Hartman).[37] The threat also, and more fundamentally, comes from the fact that the model of human selfhood claimed by Douglass is predicated on taxonomic discreteness from and ontological superiority over the nonhuman world. As Aimé Césaire, Giorgio Agamben, and Gayatri Chakravorty Spivak (among others) have argued, slavery is only the most pernicious instance of a more prevalent master–slave dichotomy that mandates a dominative relation toward everything deemed to be less than human.[38] Consequently, it may behoove us to denaturalize the category of the master as well as that of the slave rather than presume that the story ends with the conversion of the latter into the former.

All of this is to say that Douglass's successful traversing of the "nothing"–"man" binary should not blind us to the persistence of its existence. The category of the nothing survives his escape because it is independent of any individual reclassification of a "nothing" into a "man." The same is true for entire racial or ethnic groups: belatedly including African Americans in the category of the human, although a right and necessary rejection of their physical and legal brutalization, does not do away with the inferior category of the nonhuman. Nor can it, so long as the notion of the human gets its meaning from being the hypostatized opposite of nothingness; being a "westernized metaphysical self" necessarily means *not* being that which "western metaphysics" has deemed incapable of Being, from racial others to nonhuman animals to inanimate places and things. A universal humanism is, by definition and historical record, a necessarily exclusionary enterprise, no matter how capacious its eventual inclusions.[39] The human—as value—must always be differentially distributed.

The epigraph adorning *My Bondage and My Freedom*'s title page is telling in this regard: "By a principle essential to christianity, a person is eternally differenced from a thing; so that the idea of a human being, necessarily excludes the idea of property in that being."[40] The superficial meaning is benign enough: being human should preclude the possibility of being enslaved. Implicitly, however, the epigraph also suggests that ownership or enslavement is the divinely ordained condition for everything that is *not* human. You either are a "man" or you are not (worthy of consideration at all). Thus, when Douglass compares his prior condition as a slave to that of the oxen he once tended—"They were property, so was I; they were to be broken, so was I"—the point is not that oxen shouldn't be owned or broken but that *he,* as a human, shouldn't be. The comparison (and the pathos it seeks to elicit) requires an innate and rightful difference between the two: what is natural for the animal is unnatural for the man, and slavery's crime is acknowledging the principle while refusing its just application—"Behold a man transformed into a brute!"[41] Douglass's autobiographies thus radically and importantly change the boundaries of what qualifies as human, but they leave unaltered the zero-sum, either–or constitution of the category itself because, as Douglass's circular argument in "The Claims of the Negro Ethnologically Considered" (1854) makes clear, "Men" are "gloriously independent" of nature and therefore "instinctively distinguish between men and brutes."[42]

Even if we wanted to hold on to the naturalness of this distinction and claim that a human should, in fact, have more rights than an ox, we might be given pause by Douglass's insistence that a dominative relationship toward the world is humanity's defining mark (part of his argument for the equality of African Americans is that "the horse . . . admits [the African American's] mastery and dominion" just as it does that of other men).[43] To make one a man, moreover, such mastery must extend not only over animals but also, and paradoxically, over other humans. It is no coincidence that Douglass's "manhood" is finally bought not by literacy or African roots but at the cost of Covey's blood (spilled as Douglass grips his throat). Nor is it accidental that the details of the fight—bodies, sweat, blood—are given in such visceral detail that it is this scene, above all others, that stands as the defining moment of *My Bondage and My Freedom,* indeed, as the very junction between Douglass's bondage and his freedom. And lest we overlook the principle, Douglass makes it the resounding culmination of the chapter's closing lines ("Hereditary bondsmen, know ye not / Who would be free, themselves must strike the first blow?") as well as a later rallying cry to violent insurrection ("Every slave-hunter who meets bloody death . . . is an argument in favor of the manhood of our race").[44]

Some critics have sanctioned this elevation of violence as being, à la Fanon, a necessary response to an inherently violent subjugation—payment in kind, as it were (Fanon, echoing Douglass, contended that "the 'thing' which has been colonized becomes man during the same process by which it frees itself . . . [through the] murderous and decisive struggle" of "decolonization").[45] Such generally roseate endorsements, however, should be tempered both by Douglass's occasionally excessive, noninstrumental recourse to physical power (after his supposedly conclusive fight with Covey, for instance, he still "at times . . . purposely aim[ed] to provoke him to an attack . . . but could never bully him to another battle") and by its chilling result, a dehumanizing disregard for the powerless: "A man, without force, is without the essential dignity of humanity. Human nature is so constituted, that it cannot *honor* a helpless man, although it can *pity* him; and even this it cannot do long, if the signs of power do not arise." Such reasoning is not far from Emerson's perverse sentiment of abolitionist self-reliance in the "Address on the Anniversary of Emancipation of the Negroes in the British West Indies" (1844): "If the black man is feeble, and . . . not on a parity with the best race, the black man must serve, and be exterminated."[46]

Though apparently contradicting Douglass's proclamations elsewhere on the necessity of charity and reform, this is the Hegelian endgame of the exclusionary model of humanity that he inhabits: a violent enforcement of the irrevocable border between the human and the nonhuman, where the former is defined by its capacity to dominate and the latter is associated with an absence of "force." The category of "the hereditary bondsman" is maintained, that is, even as its criteria are no longer racially determined; Douglass severs the link between African ancestry and slave status, but he does not abjure the idea of natural servitude as such. This is because bondsmen who are "without force," lack "signs of power," or cannot "strike the first blow" are, by

Douglass's logic, not fully human. They are instead in an uncanny parallel to pro-slavery discourse, intrinsically property—mere things without volition or autonomy, like oxen (who, paradoxically, still need to be broken). That Douglass once was forcefully kept on the wrong side of this border does not change his attitude after he crosses over it. Rather, his success underwrites his claim that physical resistance guarantees one's status as a human subject, whereas inaction proves one's existence as a thing. To be fully human, in short, slaves must become masters. We can understand why this would be—why, having attained his humanity, Douglass is so quick to deny it to others—if we acknowledge, as James Evans has argued approvingly, that "the chief means" of "the Afro-American . . . rebellion . . . are the weapons of the oppressor."[47] Announcing his case for belonging through a nationalist and humanist discourse of rights, an individualist conception of selfhood, and a socially endorsed, masculinizing violence, Douglass expands but does not overturn the governing cultural and philosophical epistemes of nineteenth-century America.

For clarity's sake, let me repeat that Douglass was not wrong to do so; he did not sell out his African heritage for ill-gotten gain. He survived and he ensured the survival of others. Once this is stipulated, however, we might ask (all the while acknowledging the privilege that makes such a question utterable), is it possible to imagine other forms of resistance to the metaphysics of slavery, ones that turn away from the telos of the individual human subject and toward the posthumanist diasporic cosmology that Douglass necessarily rejects? Echoing the provocative formulation of John Salliant, might there yet be promise in returning to various "paths not taken in North American abolitionism"?[48]

SILENT BOOKS, TALKING SUBJECTS

As skeptical as I am about the prospect of discovering an Africanist critique of the master self within the slave autobiographies of Douglass and others, I remain compelled by the possibility of such a critique being offered by diasporic literatures. In other words, although I am critical of Gilroy's choice of genre, my own project is indebted to his (and others') example of looking to African American literature for a perspective through which to question conceptions of Being that informed the slave era—and arguably continue to inform our own. This is not to say that slave narratives do not register aspects of the diasporic cosmology. It is only to note that the imperatives of the genre—proving the autobiographical subject's rationality, Christianity, and humanity—necessitate that such elements not abide for long: they are to be considered only as they are dismissed.[49]

The early chapters of Olaudah Equiano's *Narrative,* for instance, favorably describe the traditional religious beliefs and practices of his Igbo birth-culture, but the book must soon renounce them to demonstrate his evolution into a "modern," Christian man—thereby proving the unnaturalness of his prior enslavement.[50] A particularly intriguing episode in this transition occurs when Equiano, newly enslaved on a Virginia plantation, sees in his master's home "a watch which hung upon the chimney." "I was quite surprised at the noise it made," Equiano writes, "and was afraid it would tell the gentleman any thing that I might do amiss." He continues,

> When I immediately after observed a picture hanging in the room, which appeared constantly to look at me, I was still more affrighted, having never seen such things as these before. At one time I thought it was something relative to magic; and not seeing it move I thought it might be some way the whites had to keep their great men when they died, and offer them libation as we used to do to our friendly spirits.[51]

A similar scene transpires after Equiano sees a later master and a friend reading books. He recalls,

> I had a great curiosity to talk to the books, as I thought they did; and so to learn how all things had a beginning: for that purpose I have often taken up a book, and have talked to it, and then put my ears to it, when alone, in hopes it would answer me; and I have been very much concerned when I found it remained silent.

Because of his wish to demonstrate his joy at being "providential[ly]" delivered to "the inestimable Benefits of Christianity," Equiano is quick to dismiss these survivals of his native belief system as "superstitions" or naive misperceptions—thus distancing himself, his new Enlightened self, from them (and the no-longer-him).[52] His *Narrative*, in other words, is concerned first and foremost with demonstrating that there are no biological or metaphysical impediments to Africans being civilized and Christianized; there is between Africans and Europeans only a process of acculturation, one measured by his book's narration of the ever-closing distance between Equiano-past and Equiano-present: the former made by slavery to be as much a nonagential object as the silent book, the latter an autonomous subject written into being through an act of self-authoring.[53]

I would like to resist Equiano's impulse to dismiss these episodes as merely illustrative moments of contrast between his originally "rude and uncultivated" condition and his eventual status as a rational Christian subject.[54] Pausing on their significance will not only serve as a means of introducing the larger themes of the next two chapters but will also allow me to make a general point about our received ways of interpreting scenes such as these. Note first that Equiano's attribution of life to watches, portraits, and books is based not in ignorance but in a worldview common to ATRs: the notion that superficially inanimate objects and places possess either supernatural powers or "human" attributes is a signal feature of many West African cosmologies, including that of the Igbo (centered in present-day Nigeria).[55] It is therefore not naïveté, despite Equiano's retrospective claims to the contrary, but a coherent belief system that leads him to think that his master's watch is capable of watching, the portrait of seeing, and the book of talking; in the culture of Equiano's origin, it is decidedly not the case that, as Gates argues, "only a subject can speak."[56]

The significance of this distinction lies in the different accounting it provides of the relations among slaves, things, and subjects: if in the classic interpretation slavery denies slaves their humanity by wrongly equating them with objects, in the cosmological perspective of ATRs, there is no absolute basis for distinguishing between subjects and objects in the first place; it becomes a question of use, perspective, and power rather than of essence. As we shall see in Hurston's descriptions of "living" waterfalls, stones, and ritual implements, when "things" potentially have the characteristics of "persons" (consciousness, volition, agency), it becomes difficult to hold on to a rigid sense of their ontological hierarchy.[57] One can never be *sure* that a thing is not alive, never know with certainty that one holds dominion over it rather than vice versa. You finally can't even know where you stop and the thing begins. And rather than dismiss this metaphysical flattening as irrational, premodern superstition, as Equiano, Douglass, Henry Bibb, William Wells Brown, and other ex-slave autobiographers do (thereby contributing to the semantic history by which *voodoo* becomes synonymous with *spurious*), we might consider its implications.[58]

At the very least, ATRs (and Igbo belief more specifically) compel us to acknowledge that there are more than the two ontological states of (human) being and (nonhuman) nothingness. Rather than the discrete, superior human subject dominating its essentially passive, lifeless world, we are here presented with persons radically open to the myriad things surrounding and pervading them, things now suffused with as much ontological weight as themselves. This doesn't mean that there are no differences between human and nonhuman in this cosmology, only that they are more horizontal than vertical, defined more by relation than by isolation. Indeed, far from disappearing, difference proliferates, becoming an internal as well as an external principle (recall the possibility of having multiple souls within a "single" person). A human, for instance, here becomes the effect, constellation, or sum of a certain set of (largely nonhuman) relations at different temporal and spatial scales, while a tree is the result of another (to use an example we will see in Chesnutt's conjure tales), but both are mutable networks of interactivity rather than static and sealed taxa of essentially distinct orders. The differences are thus emergent rather than pregiven, and they are capable of surprising intimacies, such that the human—now immanently heterogeneous—can no longer serve as the sole standard of being. Consequently, from the vantage of ATRs and their diasporic offspring, the slave's transition from thing to master participates, if from the opposite end, in the same exclusionary mystification that distinguished between the white human and the nonwhite–nonhuman in the first place: the delusion of slavery lies in the humanization of the colonizer as well as in the dehumanization of the colonized; it is the category of the ontologically privileged-and-transcendent human itself—born at the same moment as the dispossessed nonhuman—that is untenable within ATRs.

One might say, therefore, that although they work toward the same end—leveling the hierarchy of being between white and black—

conventional abolitionism represents the conceptual inverse to the diasporic cosmology: the former extends liberal humanist subjectivity to those previously denied it; the latter denies it to slave and slave owner alike, even as it locates supposedly "human" qualities in a range of "nonhuman" sites. Where abolitionism decries the injustice of subjugating fellow humans, the diasporic cosmology delegitimizes the naturalization of any enslavement, including that of the vital nonhuman world.[59] From this perspective, Césaire's famous "equation" that "colonization = 'thingification'" and Harriet Beecher Stowe's original subtitle for *Uncle Tom's Cabin (The Man Who Was a Thing)* rightly challenge the unnatural inequalities of colonialism and chattel slavery but fail to see that equality cannot be equal to its name so long as it is based in a violently exclusive humanism rather than in a universalization of both thingness *and* agency.[60] They do not accept that the world is more "human" than we presumed and the human, paradoxically, less so. Conversely, as Leonard E. Barrett notes, an African-derived concept of the vitality of things closes the gap between "us" and "them":

> The Black man in New World societies was reduced to the status of a thing—and a disposable thing at that, one to be used and abused at will. . . . But how much potential life is there in a "thing"? The concept of thingness has a special place in the traditional African world view. The word in the Bantu language is *kintu*, "thing." But kintu also represents . . . "agencies of life." Now, in African thought, kintu, or "things," are not merely inert; they have latent powers.[61]

Redistributing "power" beyond the human to pluralistic "agencies of life"—agencies found even in inorganic things—this vision offers a model of nonpersonhood that defies the one violently instantiated through slavery. Here the nonperson is a radically open category, encompassing humans as well as things in a common network of "potential life" and, equally importantly, potential "abuse." In so doing, the cosmology tentatively answers in the affirmative a question recently posed by Stephen Best: can we bear witness to the violence that silenced slaves without the telos of restoring their fractured voices to a dubious category of selfhood?[62]

Again, this is not to minimize slavery's horrific dehumanization of slaves, nor is it to deny the necessity of their resisting this brutalization by asserting their humanity; it is only to note that, according to the tenets of an available diasporic metaphysics, other "rituals of resistance" are possible, other forms of relative equality, ones that reject rather than adopt slave owners' fraught conceptions of humanity.[63] Here abolition comes *through* rather than *against* the foci of nonmastery, vulnerability, and impersonality. Consequently, humanism is censured less for the hypocritical, racist exclusions of its pseudouniversalism—a critique wherein its underlying truth is perceived as a given (but never duly received)—than for its fatally hierarchical division of human from non-human. Based primarily in cosmological rather than political first principles, this vision places self and world in an unending oscillation between subject and object, identity and difference. Such an oscillation, to repeat, does not preclude that inequalities in power or agency or composition may still exist, only that they not be reduced to an existentialist 1–0 binary of existent versus nonexistent, meaningful versus meaningless, to warrant the violent instrumentalization of the so-called no-thing.

LOSING ONESELF TO THE WORLD

If this lessening of subject–object distinctions could not be long entertained by slaves seeking to demonstrate their subjectivity, what was its fate in the wake of the Civil War, when the dehumanization of African Americans was attempted by other means? The next two chapters address this question at length via the works of Chesnutt and Hurston, but in the remainder of this chapter, I want to establish the contexts in which they wrote, particularly in regard to the predominant cultural attitudes toward conjure–hoodoo–voodoo in the period spanning Reconstruction to the onset of World War II. It is a difficult task, not least because of shifting attitudes and complex inter- and *intra*-racial dynamics.

That being said, for many white Americans on either side of the fin de siècle, this cosmology was regarded as evidence of African Americans' innate cultural, if not racial, backwardness, a social atavism against which the reconstituted nation could define its own progressive modernity. An 1898 article in the *San Francisco Call*

("Queer Practices of the Negro Voodoo Doctors") was typical in its diagnosis of the cultural import of "the colored race . . . patronizing 'voodoo doctors' and squandering money on 'conjuring spells'": "Civilization would appear to have little or no effect in deterring" these practices among the "negroes."[64] Primitiveness, in other words, is ahistorical; reform is for naught. Though often secular, such judgments were amenable to long-standing convictions regarding the inherent evil of African-derived spiritualities (especially voodoo and its assumed associations with cannibalism, child sacrifice, and ritual orgies), and in subsequent decades, this combination joined with American imperialist sentiment to provide a justification for the U.S. occupation of Haiti (1915–34).[65] A 1920 *New York Tribune* article titled "The Evil Spell of the Voodoo Is over Hayti" gives a sense of the prevailing rhetoric, trumpeting the marines' "task of subduing [the] barbaric sorcery" of voodoo, labeled "A Souvenir of Africa" and a "savage tribal custom." "There remains but one thing to start the Haytian native on his upward course," the article continues: "the industry which the white man can bring to this little republic."[66] Haitians are here at least granted the possibility of an "upward course" out of "barbari[sm]," but, as in the racist evolutionary historiographies discussed in chapter 2, it's a paternalistic concession that assumes the white man's supervision over the eradication of what an earlier article describes as "A Fragment of Darkest Africa on the Western Hemisphere."[67] As we shall see in the next chapter, Chesnutt, though renouncing the racial essentializations of this logic (he is careful to note that "relics of ancestral barbarism are found among all peoples"), largely shares the conviction that conjure and related practices are pernicious "superstition[s]" that need to be relinquished for black Americans to assume their proper place within "advanced civilization."[68]

By 1930, however, a marked shift had occurred in at least some people's attitudes toward voodoo. Against the backdrop of the Harlem Renaissance and increasing anti-imperialist sentiment, diasporic spirituality in general and voodoo in particular became objects of sincere religious, anthropological, and/or aesthetic inquiry for black and white persons alike.[69] Melville J. Herskovits's 1932 Boasian ethnography, *Life in a Haitian Valley*, for example, presented voodoo as a syncretic religion more closely aligned to Christianity than not, thereby countering sensationalistic, exoticized accounts such as John Craige's *Black Bagdad: The Arabian Nights Adventures of a Marine Captain in Haiti,* Richard Loederer's *Voodoo Fire in Haiti,* and W. B. Seabrook's *The Magic Island,* which popularized the term *zombie* in the United States.[70] And throughout the 1930s and 1940s, the Works Progress Administration (WPA) made a concerted effort to record and document African American folk beliefs and practices that were believed to be under threat of disappearing, thereby laying the foundation for the better-known participant–observer studies of Haitian Vodou by Hurston *(Tell My Horse: Voodoo and Life in Haiti and Jamaica),* Katherine Dunham ("The Dances of Haiti: Their Social Organization, Classification, Form, and Function"), and Maya Deren *(Divine Horsemen: The Living Gods of Haiti).*[71]

Nevertheless, even in many positive representations of African diasporic spirituality, its primitiveness still was assumed, its supposed atavism merely remade as a sensual attraction rather than a diabolical threat. In the most common version of this rhetoric, "that branch of white civilization which has been exposed to the Puritan miasma" was enjoined to

> revel in a primitive folk . . . still happy in the joyous rhythms and rich colors and free forms that have been vanishing, alas, out of our less primitive white society. . . . Suppose 135th Street . . . does look like a very short step away from the Congo? Think of the kick our own rarer, truly civilized spirits get therefrom! Esthetically envisaged, Harlem is being glorified, not libeled.[72]

Here the exoticism of black culture is not a disease to be exterminated but a cure to be disseminated, not the archaism that modernity must surpass but a sensual wellspring that must be revisited—"a white man's holiday." Through contact with the African "rhythms . . . colors . . . and forms" that survive in African American culture, "white society" may be revitalized out of its deadening hypercivilization—or at least get a temporary kick before returning to the sober business of decent living.[73] It's thus a good thing that white people, too, prove susceptible to voodoo's "powerful effect . . . on the nervous system"; in this way, they can therapeutically explore the dark continent of their own unconscious: "Hidden depths of primitive emotion of which you were unaware are roused; momentarily you become a savage. . . . The primitive lies dormant in all of us."[74] Black ids, white masks.

Such ideas often were voiced by northern white intellectuals seeking a revivifying encounter with otherness (as was the case with Hurston's early patron, Charlotte Osgood Mason), but postbellum African Americans, if for different reasons, also explored the possibility of reconstructing themselves through the "occult powers" of Africa.[75] Pauline Hopkins's *Of One Blood* (1902–3) provides a good, early example. In the novel (originally serialized in *Colored American Magazine*), a poor, ostensibly white Harvard student (Reuel Briggs) experiences a series of extraordinary paranormal events—including the awakening of his latent "power of second sight" and control over the "secret[s] of Nature"—on the way to discovering that he is in fact "descended from a race of African Kings," indeed, that he is the rightful monarch of a long-lost but advanced Ethiopian civilization to which he must "return and restore the former glory of the race."[76] Here, then, the African "wisdom and science" commonly known in part as "voodoo" enables a celebration of the "inborn dignity" of "the Negro race."[77] The transvaluation of voodoo equals the elevation of its people. And as the race is confirmed corporately, so is Briggs individually, finding through his "supernatural powers"—in a direct citation of William James—his "hidden self" (the subtitle of Hopkins's text).[78] Voodoo is therefore essentially self-affirming in *Of One Blood*, allowing "one's [false] identity . . . [to] be . . . lost . . . [while] one still . . . liv[es]" so that one's true, "ever-living . . . Ego" might be realized.[79]

Given such language, it's not surprising that a significant portion of the criticism on the novel has focused on Hopkins's engagement with black nationalist and Ethiopianist discourse, on one hand, and with the new, universalist theories of psychology, on the other.[80] If, however, the extent of Hopkins's investments in these discourses remains a debated issue, the fact that voodoo works in the novel toward the realization of holistic African American identities appears settled. And in this Hopkins anticipates not only the Harlem Renaissance's exploration of diasporic beliefs but also many current analyses of the therapeutic role of voodoo and conjure in African American culture, where these practices are said to "affir[m] personhood and hea[l] persons," facilitate a "process of self-discovery," and allow "individuals [to] actualiz[e] themselves."[81] Rather than "destr[oy] the integrity of the self," as did slavery and Jim Crow, such traditions are seen to "provid[e] an increased scope for [the self's] fulfillment."[82] Supported by much African American literature from this and later periods, these readings honor the myriad ways in which attempts to deny the history and humanity of people of African origin—often via demonizations of their beliefs—were met and defied.

In other period texts that explore this cosmology, however, we can see an alternative critique, one that disavows rather than adopts the ideal of autonomous selfhood. In this respect, it is telling that Briggs's powers owe more to Western supernatural traditions than to diasporic ones ("animal magnetism" is referenced repeatedly), and it could be argued that *Of One Blood*'s *Bildung* teleology, like that of Equiano's and Douglass's narratives, challenges exclusionary Euro-American models of subjectivity only insofar as it extends them.[83] For other pre–World War II diaspora-inspired texts, though, the spiritual and material unification of persons with "African" forces is not self-affirming; it is, rather, destabilizing because it erodes the constitutive barrier separating "us" from the world. Instead of our mastery extending over the universe, this integration entails our conscription under the dominion of powers other than ourselves—powers that constitute and exceed "us." Again, this is not to trivialize the many ways in which conjure and voodoo have been used for personal ends, from antebellum poisonings to the love charms of mail-order "African magic" catalogs in the 1930s; it is only to note, as will become evident in the Chesnutt and Hurston chapters, that the conjure cosmology is irreducible to such ends—and by the very logic of its action: the interlineation of person and environment required for conjure's agency necessarily deconstructs the autonomy, interiority, and coherence of normative personhood. To mentally or physically influence another by transcorporeal means is to be opened to the same susceptibility. What's more, there is always the very real possibility that what we take to be our use of diasporic energies is actually an instance of our being used by them. In this sense, though the diasporic cosmology does not mandate extinction—as do those of Poe and Adams—the threat of individual bodily death immanent to voodoo dolls and black cat rituals betokens a universal destruction of the category of the master self and its dominating, ontologically purifying remove.

No wonder most ex-slave narratives, committed as they are to demonstrating that their authors are "eternally differenced from . . . thing[s]," do not center themselves on this cosmology. As we have seen, aspects of it may momentarily obtain—recall the strained tense of Douglass's "I WAS A MAN NOW," a fractured temporality that labors to announce the transition from silent "thing" to literate "human"—but only invariably to be muted so that the autobiographical subject may speak himself into existence: we read Equiano, not the silent book, Douglass, not Sandy's root; inevitably, the emphasis for both authors is more on self-possession than on the possibility of being spiritually possessed. Theirs is a different agenda, of course, one necessary for the attainment of equal status, but one thereby made incapable of criticizing that status *as such*.

Yet if slave narratives must disavow the diaspora cosmology to prove their authors' humanity, as indicated by certain kinds of mastery (of writing, of violence, of others), then two historically related late-nineteenth- to early-twentieth-century genres—the African American folktale and the ethnographic study—take it as their frequent subject.[84] Developing as professional ventures at the same historical moment, the former has been recognized as including significant "survivals" from African traditions, and the latter, despite its entrenched cultural biases, at least methodologically attempted to record cultures as they were "found."[85] And as Michael A. Elliott, Brad Evans, and Bill Brown have shown, this shared concern with factual, scientific, or objective representation led to extensive mutual influence between the two, indicating, like Poe's earlier texts and Adams's contemporaneous ones, the extent to which our modern disciplinary distinctions had yet to rigidify.[86]

My specific texts are Chesnutt's conjure tales (heavily drawn from black American folk idioms, oral narratives, and worldviews) and Hurston's participant–observer explorations of southern U.S. and Caribbean black cultures in *Mules and Men* and *Tell My Horse*. I focus on these texts in particular because they present the conjure universe in such a way as to underscore its direct repudiation of the master–slave ontology in both its racial and its anthropocentric guises. Admittedly, Chesnutt's and Hurston's texts occasionally withdraw from or buckle under the force of what they are presenting; at times, they retreat into frame narratives (Chesnutt) or abruptly end at the moment of greatest contact (Hurston). We might then ask, why approach diasporic spiritualities through narrative at all? Precisely because it is in narrative—supposedly the paradigmatic expression of Western subjectivity—that their consequences for a particular conception of Being are lit up, where the torsional effects of nonhuman forces on human selves are made most clear. Thus, even—indeed, precisely—when Chesnutt's and Hurston's texts break down or attempt to escape their subjects, they manifest the posthumanist challenge that the cosmology represents.[87]

The challenge extends beyond the texts as well, of course, having at least two distinct antipodes within turn-of-the-century to mid-twentieth-century American culture, roughly the period between the publication of Chesnutt's and Hurston's major works. The first is the notion of an essential divide between human and nonhuman, discussed earlier in the context of hereditary slavery but which continued in postbellum America to equate African Americans with something less than white humanity (most visible in Jim Crow and *Plessy*'s lie of separate equality). Equally under pressure are the historical correlatives of these exclusions: capitalistic and religious individualism (each perhaps best captured in Andrew Carnegie's 1889 *Gospel of Wealth*, published two years after Chesnutt's first story in the *Atlantic Monthly*) and manifest destiny's exploitative relation toward the natural world (as evidenced by the new industrial-scale logging and mining operations of the late nineteenth century).

The second antithesis of this cosmology appears to be the inverse of the first. Rightly understood, however, it amounts to the same thing. It is, in short, what might be called the West's *non sum qualis eram* recoil against the excesses of its own Enlightenment heritage, and it is characterized by the putative belief that the individual human subject is affirmed and empowered by its integration within a larger, holistic network of being. The enthusiasm directed toward diasporic spiritualities during the period is one manifestation of this phenomenon, but related examples would be other non- or para-Christian beliefs, such as spiritualism, New Thought, and Theosophy, that celebrated the self's enfolding within the cosmic One[88] and early environmental conservation and preservation movements that appealed to a similarly self-affirming conjunction.

Though these diverse discourses frequently advocate abandoning individualistic selfhood and its inherently self-interested relationship to the world, they are yet structured by an anthropocentric logic of self-gratification that fundamentally mirrors it. Fusing with the cosmos means merely assimilating its power into oneself; far from being abdicated, the self is universalized. Both John Muir and John Burroughs are thus typical in suggesting that the sublime apprehension of the self's minuteness with respect to the world becomes a paradoxical means of cosmic elevation.[89] Similarly, Madame Blavatsky's widely read and highly influential Theosophical text *The Secret Doctrine* argues that the path to enlightenment lies in "the martyrdom of self-conscious existence," or, more simply, "sacrificing" oneself, but then paradoxically claims that "it is the HIGHER-SELF, the real EGO who alone is divine and GOD."[90] Therefore, like that which they protest—the Enlightenment's self–world binary, Protestant personal salvation, capitalistic self-advancement, etc.—these worldviews betray an ultimately narcissistic relationship to the cosmos, seeing in it only a means to realize the self's full potential. Harmonizing with the universe becomes only another colonization.

For these dual models of subjectivity that mistake the world as either wholly other or wholly self, the universe announced in Chesnutt's and Hurston's texts presents an insolvable crisis because it refuses to locate an abyss between the personal "we" and the impersonal "it." Again, this doesn't mean that the human and the self don't exist, only that their existence is defined by interdependence rather than independence, by milieus rather than vacuums, defined, that is, by the presence, not absence, of connections, by (rhizomatic) *rootedness*—in all senses of the word (both Chesnutt's stories and Hurston's studies ask us to consider how, in the presence of rootwork's transbodily influence, a person can be deemed discrete, a self-contained free agent autonomous from its environment, when we wouldn't think the same of a tree). Neither the cosmos nor outside of it, "neither creator nor ultimate controller," "we" are in the most fundamental sense like every other thing: compositions of relations that give rise to, but cannot be bounded by, "us"—temporary closures that emerge and disappear according to alien-yet-immanent processes.[91] In the language of a Chesnutt tale, "we" are a muddy eddy of strange tides, or, in an image of Hurston's, "we" are an altar of difference, an assemblage of unlike bits, pieces, and blood that conjures an ephemeral identity-in-action. Consequently, the networks of our (un)becoming cannot easily be reduced to discriminations of sameness versus otherness, self versus not-self.

The next chapter explores this attenuation of the "I"–world binary in relation to Chesnutt's supernatural plantation tales, which picture persons being transformed into aspects of their nonhuman environments—frequently to fatal effect. In this, the stories, despite their frame narratives' best efforts, reveal the impossibility of a humanist conjure, thereby anticipating recent posthumanist or "dark" ecocriticisms that replace early-twentieth-century visions of idyllic ecological equilibrium (which assumed the separation of human from nonhuman to balance them) with ones of unsettled (and unsettling) turbulence. That these stories' equation of conjure with death is a function of their antebellum setting rather than an innate aspect of conjure itself is illustrated in the subsequent chapter on Hurston. Death is still a real threat in her hoodoo–voodoo ethnographies, but she makes clear that it is only one form of the humanist self's unmaking. More common is the breaking of this self's detached rationality and assured control on contacting powers it can only partially comprehend, a limit defined not by ineffable distance but by fearful intimacy. Such a disassembly resonates with phenomenologies of religion in both Hurston's age and our own, but I will argue that it ultimately goes beyond the anthropocentrism of many of these accounts by revealing the emergent entity of the self—its identity and difference, life and death, agency and passivity—to be indissociable from its material–spiritual contexts.

The common challenge thus proposed by Chesnutt's and Hurston's cosmologies, born anew in the crucible of New World slavery, is what to do with our selves when we can neither own them nor give them away. It is a question that arises in the conjure cosmos because here, unlike in Poe's and Adams's respective universes, life goes on even if "we"—masters of our selves and of our world—don't. "We" may be dethroned as lords over creation, but a transformed, heterogeneous version of "us" may yet survive. Stripped of the easy finalities of extinction and immortality, determinism and free will, a suddenly impure, less-and-more-than-"we" must face the prospect of continuing to exist in a universe that makes "us" both possessed and possessors, causes and effects.

Is there a politics or ethics adequate to this interstitial condition? What would it mean to stipulate the existence of the human but suspend the certainty of its ontological remove, not only from the "higher" mammals but also from matter, time, and space as such? Does a flat ontology require that we afford an ox, a spirit, a watch, a painting, or a book the same rights as those denied the slave? Can cosmic horizontality establish a "democracy of objects," a "parliament of things," or a "cosmopolitics" without either reasserting our mastery (even in the form of stewardship) or anthropomorphizing the world (even for its protection)?[92] Or will it instead make dehumanized slaves of us all, as Barbara Christian and bell hooks once complained of post-structuralism's denuding of subjectivity in the wake of the civil rights movement, when personhood was finally fully recognized in those historically denied it?[93] Is it possible to live an acknowledgment that "we"—however broadly defined—have some agency but not all, and that what we do possess is not entirely our own? These questions will get a fuller hearing in the coda, but we can note here that it is at least conceivable that the becoming posthuman of both persons and world cannot be domesticated by our ethicopolitical attempts to legislate it.

Yet this need not be the end of the story: in moving beyond the zero-sum notion of rights based in the ontological scarcity of the anthropocentric standard—and possibly beyond the discourse of rights altogether—we might discover the potential of the conjure cosmos's radical antipolitics. Cognizant of its dialectical resonances, I say "antipolitics" because its imperative is neither to extend nor to abandon the category of the human but to recognize that the human is only one assemblage (itself composite) among others so infinite in kind, scale, duration, and combination that they can be acknowledged but their meanings—their metaphysical weight—never fully known or instrumentalized (whether for violence or tolerance). It is here that ANT and the diaspora cosmology may be put into most productive conversation, as both stipulate what we might otherwise reject as ignorance or unscientific superstition: the animistic possibility that agency and vitality are abundant rather than scarce, circulatory rather than fixed, and in between beings rather than properties of beings-in-themselves.[94] As a result, the “individual”—at whatever scale of definition—is made a stochastic transaction, an uncertain negotiation, an unruly gathering, a variable resonance; the “I,” the “other,” and the “you” as amalgam, as IOU; or, in the language of Chesnutt’s and Hurston’s texts, self and not-self as impossibly imbricated co-conjurers bodied forth in a common flesh or clay. For both ANT and hoodoo, then, we can never be certain that what is before “us” is not, potentially, one of “us”—or “we” part of “it”—because all are defined by the simultaneously constitutive and estranging *process* of their relation. Such a formal equality—one of both power and susceptibility—grants that humans and nonhumans are different without giving that difference a fixed content or value. Consequently, however, our response to this indeterminacy must itself be radically underdetermined: it may defrock the murderous certainty of our orders, but it cannot tell us which new habits to assume.

Necessarily lacking the slave autobiography’s overtly political agenda, this perspective may seem woefully inadequate—even disabling—given our present circumstance. But in refusing to assume that the only way out is through, by questioning rather than intensifying those prerogatives of the individual subject (its autonomy, property, priority, mastery, etc.) that arguably undergird the racist and anthropocentric dominations of our times, it at least offers another way of imagining the future, even as it leaves “our” place within it uncertain. Perhaps, therefore, like Chesnutt and Hurston, we can only approach this vision inevitably to withdraw from it, never fully apprehending it lest we lose our selves. Or perhaps we may sense, fleetingly, that such a loss might already be.

4

HOODOO YOU THINK YOU ARE?
Self-Conjuration in Chesnutt's The Conjure Woman

"MOST OF THE DELUSIONS connected with this belief in conjuration grow out of mere lack of enlightenment. . . . Relics of ancestral barbarism are found among all peoples, but advanced civilization has at least shaken off the more obvious absurdities of superstition."[1] This is Charles Chesnutt's assessment of conjure in his 1901 article "Superstitions and Folklore of the South," and given the fact that he had published *The Conjure Woman* only two years earlier, it's hard not to see the irony: Chesnutt did not believe in the beliefs that he helped to popularize, indeed, that gave him an opportunity to become a published author in the first place. Aspiring to be a writer from an early age and continuing to pursue the ambition even as he became a successful stenographer, it was his conjure tales—strategically calculated to capitalize on the overwhelming popularity of Joel Chandler Harris's *Uncle Remus* stories and Albion Tourgée's *A Fool's Errand*—that provided his entry into the literary marketplace. Yet despite conceding that "these vanishing traditions might furnish valuable data for the sociologist . . . in the future study of racial development" because they "probably grew . . . out of African fetichism," Chesnutt consistently refused to acknowledge anything more than "goopher['s]" ethnographic worth. Far from being a coherent belief system in its own right, an essential component of African American identity, or a symptom of an innate racial susceptibility to superstition, such practices were but fading vestiges of an artificially sustained archaism, of interest only as analogues to the early stages of "the white man's civilization." Chesnutt, in other words, offers an evolutionary cultural historiography without the entrenched racism of the kind seen in chapter 2, meaning that anyone testifying belief in conjure was not biologically inferior but merely "lying," a "fraud," or "self-deluded."[2]

Some have claimed that this dismissal of conjuration was only a ruse by which Chesnutt defused white skepticism toward a cultural tradition he actually respected, that he was only "signifying" on "the voice of contemporary ethnography while working inside it to preserve African American cultural forms."[3] We will see in the next chapter that something like this is true of Zora Neale Hurston's anthropological studies of hoodoo and voodoo. It is difficult to be convinced that the same holds for Chesnutt, however, especially given the consistency of his denunciations of conjure and the fact that, according to Richard Brodhead, he "spent his early career trying to teach black children out of dialect speech and folk superstition."[4] Indeed, Chesnutt modeled his entire life after the kind of "enlightenment" principles—Gilded Age entrepreneurialism, individualist self-determination, rigorous self-improvement—that he believed would, if extended to the rural masses, abolish the patent "absurdities" of black folk culture. Like Frederick Douglass before him, and many of his educated African American contemporaries, he powerfully denounced the hypocritical, racist exclusions of these "universal" principles but never the principles themselves, never advocated for their abandonment in favor of diasporic beliefs that only survived, he suggested, because of the brutal denial of equal education to African Americans in both the antebellum and post-Reconstruction South.[5]

Of course, belief in one's sources was by no means requisite for the kind of writing that Chesnutt did early in his career. Readers of regionalist and plantation fiction at the turn of the century assumed that authors would possess enough familiarity with the cultures they presented to make them authentic, but equally important, they also expected a certain amount of authorial distance from them, a national coloring on a local theme. Had Chesnutt actually professed faith in conjure, it most likely would have precluded his fiction from appearing in magazines such as the *Atlantic Monthly.* As Jeffrey E. Anderson's *Conjure in African American Society* indicates, white interest in black folk beliefs spiked at the turn of the century, but mostly as a means to "bolster white superiority and regional distinctiveness" in the face of homogenizing industrial forces.[6] Typical attitudes toward conjure thus ranged from accusations of demonic influence to paternalistic preservationism to casual linkages of African Americans with a primitive not-quite-past, but all postulated the practice's—and practitioners'—essential otherness (drawing on all three impulses, a representative article from 1890 argues that "negroes, living at a convenient distance from churches, schools, and railroads, are found to have as firm a belief in witchcraft as those savages of the African bush who file their teeth and perforate the cartilage of their noses").[7]

It's safe to assume, then, that it was the combination of Chesnutt's personal, "childish recollection" of the "old-time belief in what [was] known as 'conjuration' or 'goopher,'" gleaned from his years as a youth in southeastern North Carolina, *and* his publicly professed skepticism toward these beliefs (note the choice of "childish" rather than "childhood") that contributed to the measure of national success he briefly enjoyed.[8] Combined with his request that Houghton Mifflin not "advertise" his race upon publication of *The Conjure Woman* lest it "affect [the] reception [un]favorably," this remove inoculated Chesnutt against what otherwise might have seemed a too-close identification with practices that, though pervasive among both white and black Southerners (and, after the war, Northerners as well), were coded in the mainstream press as either essential or conditional marks of African Americans' spiritual, cultural, and intellectual inferiority.[9]

We need not look to Chesnutt's subsequent writings, though, to see his marked distance from the folk beliefs *The Conjure Woman* presents. The tales themselves are framed, as it were, by suspicion: set in Reconstruction-era North Carolina, each story has the white carpetbagger John introduce one of ex-slave Julius's supernatural folktales of antebellum plantation life and, once completed, dismiss it as a ruse by which Julius, now working as John's coachman, attempts to achieve some ulterior motive. As invariably as Julius proclaims the truth of his magical tales, John protests it, seeing an instrumental, this-worldly incentive behind Julius's performances. Admittedly, as readers have long noticed, the stories implicitly criticize John for projecting his own hyperrational, profit-driven view of the world onto Julius in this way. Unlike his wife, Annie, John consistently fails to appreciate the actual or symbolic truth of Julius's accounts of conjure offering slaves an alternative power with which to meet, if never equally or adequately, the brutalizations of enslavement.[10]

But what critics are less quick to acknowledge is that Chesnutt's criticism cuts both ways, toward Julius's beliefs as well as John's unsympathetic dismissal of them. Julius, for instance, is shown to be variously uneducated—"dey's a young nigger gwine ter school in town, en he come out heah de yuther day en 'lowed dat de sun stood still en de yeath turnt roun' eve'y day on a kinder axletree. I tol' dat young nigger ef he didn' take hisse'f 'way wid dem lies, I'd take a buggy-trace ter 'im"—and comically superstitious: regarding the lucky "rabbit foot" he has owned for forty years, Julius explains, "De fo'-foot ain' got no power. It has ter be de hin'-foot, suh,—de lef' hin'-foot er a grabe-ya'd rabbit, killt by a cross-eyed nigger on a da'k night in de full er de moon."[11] Although we might want to ascribe these instances to an elaborate performance designed to meet and exploit white expectations, John's feeling about such beliefs anticipates almost exactly Chesnutt's in "Superstitions and Folklore of the South": "Your people will never rise in the world until they throw off these childish superstitions and learn to live by the light of reason and common sense." I think, then, that we are meant to share in the humor of John's dryly sarcastic response to Julius's exacting prescription regarding a "true" rabbit's foot: "They must be very rare and valuable."[12]

This is only to note that the predominant tendency in the criticism to align Chesnutt (and thus itself) with Julius and his folk beliefs rather than with John and his rationality is not as straightforward a proposition as it might at first appear. As Brodhead notes, "Chesnutt himself was considerably closer to John's position than to Julius's. To identify Chesnutt with Julius's black vernacular, as is commonly done, is to fall into this genre's trap of identifying all blacks with a single image of black culture."[13] The larger point I want to make, however, involves the curious thing that happens to conjure when Julius is made Chesnutt's proxy. Eric Sundquist's influential *To Wake the Nations: Race in the Making of American Literature* offers one of the most compelling examples. Sundquist argues that the conjure of *The Conjure Woman* should be understood as operating on three mutually supportive levels: that of Chesnutt, that of Julius, and that of Aunt Peggy, the primary conjurer in most of the tales. Although only the last is associated directly with actual conjure practices, Sundquist can claim that the other two equally embody the essence of conjure by brilliantly transforming the medium of its action from the physical to the rhetorical. Conjure's metamorphic powers become, via Sundquist's alchemy, the bewitching verbal performances through which Chesnutt and Julius fashion themselves and beguile others; Peggy's material art becomes Chesnutt's and Julius's signifying Art ("conjure . . . transfigured into a narrative strategy"), with both importantly deriving from African origins: "Like the conjure woman working her roots," Sundquist writes, "Chesnutt employs the power of linguistic metamorphosis to work his own 'roots' . . . [thereby] recovering the past of slave culture and beyond it echoes . . . of the middle passage and African life."[14]

For Sundquist, the significance of this recovery of an African signifying practice analogous to an African conjure practice is that it provides Chesnutt with an extrinsic, diasporic space–time from which to critique the continuity between U.S. ante- and postbellum racisms (including, potentially, that of his own audience)—all while not appearing to do so, indeed, even while appearing to be complicit with it (many of the stories rely on such stereotypes as "Why the Darkey Loves Chicken"). "Chesnutt," Sundquist contends, "carried the trope of conjure, if not a personal belief in it, into the very stratagem of his own career. He wore the necessary mask of whiteness . . . but he claimed his African American heritage in promoting the authenticity of his literary voice."[15] Turning the assumption of his race's inequality against itself, Chesnutt tricks his readers with a show of conformity that disguises its own dismantling. And the same general point holds true for Julius's stories, which "wor[k] by indirection and 'nonsense'": they subversively signify on the stereotypes of racial inferiority that they ostensibly perform, offering an act of covert resistance "depend[ent]" on "the figurative power of conjure" as an "allegor[y]" of a forceful yet (almost) invisible oppositional power.[16] Both Chesnutt and Julius thus denaturalize the racist expectations of their respective white audiences by enacting them—with a self-empowering difference. In so doing, Sundquist concludes, they are quintessential examples of the collective "voice of African American culture," which is "itself a voice of metamorphosis and transformation."[17]

I cite Sundquist to this extent because of the power and influence of his account, which has rightly become a touchstone for the past twenty years of Chesnutt studies. Indeed, the conjure–signifying pairing is now paradigmatic.[18] Happily, it is also largely correct. Julius's tales do invoke the slave past as a way of reflecting on the lasting racism of Chesnutt's contemporary moment, including its at least partial embodiment in John (and his dismissive condescension toward Julius). Moreover, Chesnutt—in a direct repudiation of the marked one-dimensionality of Harris's Uncle Remus—endows his black characters with a capacity to frustrate those whites who, based on fallacious notions of their own superiority, would presume to exploit them. And the agency of this resistance for Julius, as for Chesnutt, is a form of storytelling that stages power inequalities as a means of subtly redressing them in the teller's interests, thereby performatively disproving the charge of African Americans' nonintellectuality ("Julius," Matthew R. Martin summarizes, "is every bit the white man's equal").[19] In this sense, conjure-as-rhetorical-performance does become a means to conjure oneself, for author and character alike.

Despite these recommendations, however, there are problems inherent to reducing conjure to a mere metaphor for signifying acts of racial affirmation. One is that it too easily lapses into exaggerated claims of Chesnutt's identification with a general "black" identity (his initial motivation for writing plantation fiction, despite its patent racism, was personal—not racial—uplift: "I want fame; I want money").[20] In fact, throughout his career, Chesnutt was more concerned with combating racism than with recuperating essentialist conceptions of blackness (his "The Future American" postulates—and eagerly anticipates—an imminent blending of races to the point of indistinction).[21] But the more significant shortcoming of the dominant critical approach lies not so much in what it adds as in what it leaves out, namely, the actual conjure practices and cosmology that the tales present. By conceiving of conjure as only a material (and less significant) instantiation of what is essentially (and more importantly) a linguistic process, Sundquist and others are quick to fold the former into the latter.[22] Conjure thus appears in these accounts only as a figure for figuration itself, as "a potent reservoir of meaning and linguistic power." Meaningful "vocalization[s]" take priority over material charms, "improvisational flexibility" over bodily metamorphosis, the spoken, in a word, over the token.[23] Consequently, the drama of the conjure tales is made indistinguishable from that of Chesnutt's subsequent novels (which largely abstain from the supernatural to focus directly on the this-worldly horrors of slavery's aftermath): it is fundamentally "psychological," involving conjure only to the extent that it is a powerful symbol for the "hidden" inner lives, "memor[ies]," and potentials of an oppressed yet defiant people.[24]

Again, such an account isn't wrong (given his aforementioned disdain for black "superstition," Chesnutt himself likely would have agreed with it), it's only incomplete, which wouldn't be a problem, except that turning conjure into an oppositional semantics invariably involves displacing it as an oppositional metaphysics—dispossessing it (so to speak) of its original force. In the present chapter, I seek to redress this omission by examining the tales' exploration of the conjure cosmology without recourse to the hermeneutics of depth or symptom. In doing so, I will insist on the differences, long ignored, between the frame narratives and the tales they introduce, between signifying language and conjure ritual, between a politics of the liberal subject and an ontology of environments. Importantly, we do not thereby lose the stories' deconstruction of racism; as we shall see, conjure, too, denatures the metaphysics of white privilege. Here, however, equality is attained not through the articulation of a "black" subjectivity structurally identical to that of "whites" but through the disarticulation of the subject–object distinction as such. This may seem an abstract, ineffectual response to the disenfranchisement, enforced sharecropping, and organized violence of Jim Crow (lynchings reached a peak in the 1890s, a fact of which Chesnutt was well aware), but it begs the question, of what worth is an equality with those who would perpetuate such crimes, of what value a status that not only countenances but *underwrites* these horrors?[25] The answer, of course, is survival. As asked in the previous chapter, though, once survival is stipulated, might there be an opportunity to reexamine rather than merely extend the dubious category of personhood that authorized the enslavement of the nonhuman world (originally designed so as to include Africans)? Might there be another way to meet its injustices? Posing these questions isn't another version of white privilege; the conjure cosmology—as illustrated in Chesnutt's tales—offers just such a rebuke.

Significantly, though, the cosmology also disavows the apparent inverse of this domineering subject position: the turn-of-the-century proto-environmentalisms of John Muir, John Burroughs, and others who sought to harmonize the individual with the sublimity of the nation's natural grandeur and, through it, the entirety of the universe. As Paul Outka has recently argued, such attempts, though less obviously violent in their effects, commonly perpetuated slavery's "core mystification" of "view[ing] African Americans as an extension of the natural world."[26] Consequently, African Americans were not only excluded from transcendental experiences with wilderness (because they could not unite with what they already were), they also were conscripted within an instrumental, anthropocentric worldview that made them merely another spiritual resource for, in Muir's words, a "tired, nerve-shaken, over-civilized people."[27]

For its part, the conjure cosmos disallows both forms of self-constitutive mastery over the nonhuman world, the provocation of Cartesian separation *and* the allure of romantic union. Turning persons into trees, animals, and clay—or confusing the self's borders with those of its voodoo doll likeness—conjure pictures transformation without transcendence, determinism without teleology, integration without elevation. Indeed, more often than not, Julius's tales show contact with conjure ending in death. Undoubtedly, Outka is right to claim that this is attributable in part to the stories' antebellum settings, with conjure's dehumanizing metamorphoses mirroring slavery's violent "conflation of racialized subjects and the nonhuman natural."[28] However, as we will see most clearly in the next chapter, death, though always present as a possibility, is not an inevitability of conjure practice (if it were, the practice would be rather short-lived). Yet the mortal risk of conjure is not merely a trope of either slavery or the death of a primitive (according to Chesnutt) belief system but a real potential that signals this universe's incompatibility with our domination of it. As in the posthumanist "dark ecology" of Timothy Morton and others, "we" are here more conquered than conquerors, possessed, not possessors. Indeed, our permeability to "external" forces—to the point of being wholly transformed in both body and mind—suggests that "we" are not discrete entities but ephemeral, hybridized confluences of impersonal, nonhuman powers. "We" exist, but we cannot finally delineate the boundary between "us" and "not-us" because "we" *are* this uncertainty, this real yet nebulous conjunction. From the conjure perspective, therefore, there is no "we" that precedes its alienation, meaning that the tales disarticulate as materially false the very identities that Chesnutt criticism celebrates as empoweringly—because verbally—self-made.

This conflict becomes most apparent when we notice the stark difference between the conjure tales proper (as told by Julius) and the frame narratives that surround them (as written by John). Like the Uncle Remus–white child frames of Harris's Africanist animal tales, the frame narratives in *The Conjure Woman* mark a humanizing circumscription around an inherently nonhuman world. Although the ideological functions of the frames of *Uncle Remus: His Songs and Sayings* and *The Conjure Woman* differ (naturalizing and denaturalizing, respectively, the New South's myth of the Old South), both serve to reify the boundary separating the relentlessly a-human "fictions" of the oral tales from what we might variously call the "human," the "personal," the "ethical," the "psychological," or the "real" of the frame world.[29] Aunt Peggy and the universe of material conjure she represents do not exist in the "real" world; Uncle Julius and his signifying language do.[30] Cordoning off the cosmic so that the psychic can be legible, we see hypostatized here the generic difference dividing synchronic folklore from diachronic realism (and the former from the ethnography to which the latter aspires).[31] Such a distinction, of course, is also one of authorship: though Chesnutt was compelled to qualify his claims about the originality of his conjure tales after recognizing the degree to which their themes and even plots were indebted to the folklore traditions he heard in his youth, he undoubtedly authored these tales' realist frame narratives—frames defined in part by their pronounced distance from the tales' conjure universe.[32]

But if conjure doesn't exist in the frame narratives, then neither do autonomous subjects in the tales. Whereas the frames are populated with what we might identify as "real," three-dimensional people, the conjure stories have few such inhabitants (interestingly, in tales like "Dave's Neckliss," which lack supernatural elements and are not based in diasporic folklore, the characters are more recognizably "human"; not coincidentally, then, "Dave's Neckliss" is the one story in which

Julius appears as both narrator and character). Again, this is certainly due in part to the tales' antebellum settings (where slaves' subhumanity was paradoxically both assumed and enforced), but it is also attributable to the metaphysical coordinates of the conjure universe itself. Compare Julius and Aunt Peggy: the kinds of humanity, subjectivity, and psychological depth that critics prize in the former simply aren't apparent in the latter. She may, for instance, demand a ham as payment for her services, but we never see her eat it, much less enjoy it, as we do with Julius. Omnipresent but nowhere fully a self, Aunt Peggy serves (as) a function: she is less a person than a mediator, the interface through which the conjure universe is accessed.[33] Thus the difference between Julius and John, essentially one of power, is less, not more, than that between Julius and Peggy—one of "existence" itself. Indeed, it can't be otherwise, at least insofar as the predicates of normative subjectivity are antithetical to the posthuman conjure principles that animate the tales.

The "dehumanization" of slavery is not, however, the same as that of conjure. This is an important point, one discussed in the previous chapter but worth repeating here. Whereas the former mystifyingly creates an inferior world to serve humanity's existence, the latter postulates a cosmos that negates the ontologically superior human and, thus, its equally fabricated, to-be-dominated other. Though both tale and frame undermine the racist assumptions regarding the inferiority of African Americans, there is within the tales a surplus of disintegrative energy that dissolves not only race but also the preeminence of the *human* race, not only the inequalities between white and black but also those between human and nonhuman, subject and object. From the perspective of the tales, the frames' undermining of racial inferiority is thus necessary but incomplete: like the political agitation of the slave narratives discussed previously, it expands but cannot abolish the concept (and the violence) of the privileged human. Put another way, if the frames are the realm of rational, conscious agents, then the tales mark the uncanny return of the repressed. This doesn't mean that humans don't exist, only that they can't predicate their existence—or their violence—on meta/physical autonomy, purity, and priority.

CHESNUTT'S FIRST MAJOR PUBLICATION, "The Goophered Grapevine," crystallizes many of these points. Originally published in the

Atlantic Monthly in 1887, subsequently selected to open 1899's *The Conjure Woman,* and now the most anthologized of Chesnutt's works, "The Goophered Grapevine" is also the one story that Chesnutt always admitted to being based on traditional folklore.[34] In Chesnutt's redaction, the tale is occasioned when John and Annie move from northern Ohio to North Carolina for a climate more suitable to Annie's "poor health" and for the fact that "labor was cheap, and land could be bought for a mere song." While surveying an abandoned vineyard that John is considering buying, they encounter Julius in the fields. Hearing John's plans, Julius cautions against the purchase, saying that the land is "goophered,—cunju'd, bewitch'," and proceeds to tell the tale that explains the curse's origin.[35]

The vineyard, according to Julius, used to be a part of the plantation of Julius's ex-master, "Mars Dugal' McAdoo" ("Goophered Grapevine," 35). At some point prior to the war, McAdoo discovers that his slaves are stealing grapes, and when he can find no other way to stop the theft, he goes to Aunt Peggy, who is known to "wuk de mos' powefulles' kin' er goopher,—could make people hab fits, er rheumatiz, er make 'em des dwinel away en die; en dey say she went out ridin' de niggers at night, fer she wuz a witch 'sides bein' a cunjuh 'oman" ("Goophered Grapevine," 36). Because of this fearsome reputation, McAdoo pays Peggy to "goopher de grapvimes" so that "a'er a nigger w'at eat dem grapes 'ud be sho ter die inside'n twel' mont's" ("Goophered Grapevine," 37). The plan works: upon seeing three people die after eating the grapes, the slaves promptly quit stealing them.[36] A problem arises, however, when a new slave, Henry, eats grapes without knowing the consequences. Not wanting to lose a man who "could do a big day's wuk" ("Goophered Grapevine," 37), the overseer takes Henry to Peggy to see if the curse can be removed. Peggy complies but says that the only treatment is for Henry to take the sap from the grapevines every spring and rub it on his head.

What follows determines the rest of the story. When Henry does as Peggy directs and applies the sap to his head,

> de beatenes' thing you eber see happen ter Henry. Up ter dat time he wuz ez ball ez a sweeten' 'tater, but des ez soon ez de young leaves begun ter come out on de grapevimes, de ha'r begun ter grow out on Henry's head, en by de middle er de summer he had

> de bigges' head er ha'r on de plantation. . . . Henry's ha'r begun to quirl all up in little balls, des like dis yer reg'lar grapy ha'r, en by de time de grapes got ripe his head look des like a bunch er grapes. . . .
>
> But dat wa'n't de quare' thing. . . . When Henry come ter de plantation, he wuz gittin' a little ole an stiff in de j'ints. But dat summer he got des ez spry en libely ez any young nigger on de plantation. . . . But de mos' cur'ouses' thing happen' in de fall, when de sap begin ter go down in de grapevines . . . de begun ter straighten out'n Henry's ha'r; en w'en de leaves begin ter fall, Henry's ha'r 'mence' ter drap out . . . en he begin ter git ole en stiff in de j'ints ag'in, en paid no mo' 'tention ter de gals dyoin' de whole winter. ("Goophered Grapevine," 39)

Quickly realizing that Henry's annual growth and decline is tied to the natural cycle of the grapevine, McAdoo decides to capitalize on the phenomenon by selling Henry at a high price in the spring, when Henry is at the apex of his strength, and then buying him back at a discount in the winter, when his body degenerates. After five years, however, this lucrative enterprise is interrupted when McAdoo becomes "bewitch' wid [a] Yankee" salesman who convinces him to try a new, modern technique of cultivation ("Goophered Grapevine," 41). Following a brief period of explosive growth, this action eventuates in the death of both the vines and Henry: "When de vimes sta'ted to wither, Henry 'mence ter complain er his rheumatiz. . . . [And] when de big vime whar he got de sap . . . turned yaller en died, Henry died too" ("Goophered Grapevine," 42). Angered at being duped, McAdoo eagerly joins the Confederate Army when war breaks out to revenge himself on some, any, Yankees—only to be killed himself.

This is by no means a simple story to interpret. But one possibility is that Julius tells it to protect his interests: he warns in closing his narrative that a few of the goophered vines still remain and that it would therefore be unwise for anyone who could not detect them to purchase the land. John ignores the advice, buys the land, and soon thereafter discovers that "Julius had occupied a cabin on the place for many years, and derived a respectable revenue from the product of the neglected grapevines." He surmises that "this, doubtless accounted for [Julius's] advice . . . not to buy the vineyard" ("Goophered Grapevine,"

43). Perhaps unsurprisingly, many critics have focused on this tangible site of conflict within the story, as Julius attempts to manipulate John, and John tries equally self-interestedly to depict Julius and his tale in exclusively economic terms. And, as these critics have argued, it seems clear that if John ultimately "wins" the encounter (he does, after all, buy the land despite Julius's warning), Julius yet proves himself capable of criticizing John's agenda (and the racial privilege that enables it), while at the same time advancing his own.

More to the point, Julius's story illustrates that "white" knowledge, despite its ruthless drive to mastery, is ultimately less effective than that of black people. In addition to the lethal ineptitude of the Yankee salesman (an interloper-figure long identified with John in the criticism), there is also "a mighty fine [white] doctor" whom one of Henry's cheated, temporary owners consults when Henry's health declines in the winter. His "med'cine," however, "didn' 'pear ter do no good.... [And when] Henry tole de doctor 'bout de goopher,... de doctor des laff at 'im" ("Goophered Grapevine," 40). In reality, there is only one true "doctor" in the story: Aunt Peggy, a black woman who successfully manipulates the natural order rather than destroying or ignoring it, as the salesman and the doctor respectively do.[37] The lesson is unmistakable: should John ignore Julius's warning (and the local, situated knowledge that it represents), he risks having McAdoo's fate become his own. Julius's story thus puts to real-world and self-interested effect the antihegemonic force that Peggy's conjuration symbolizes.

This reading compellingly enunciates the story's censure of McAdoo's racist assumption that an ontological divide separates him from his slaves, an assumption best captured in his heartless equation of Henry and the vines as merely lost property ("Mars' Dugal' tuk on might'ly 'bout losin' his vimes en his nigger in de same year" ["Goophered Grapevine," 42]). As Jeffrey Myers notes, "by tying Henry's condition to the grapevines... the narrator symbolically portrays the dehumanization that slavery entailed."[38] And by underscoring the disastrous failure of McAdoo's attempt to master person and plantation alike, and then implicitly warning John of a similar end, Julius can be seen to deconstruct the ongoing metaphysics of racial difference in his own era.[39]

But the story's deconstructive force need not result from Julius's rhetorical performance alone, nor need it stop at the destabilization of race. The tale suggests that conjure, too, lights up the insufficiencies of white knowledge–power and that it does so by canceling the Enlightenment subject's constitutive remove from its environment, thereby exploding not only the white–black binary but the self–world one as well. As the story makes clear, Henry's identity is physically and metaphysically prosthetic with the landscape, meaning that his person is transactional rather than essential, open rather than closed. Much like a grapevine, he is an emergent, ephemeral event of powers beyond his control yet immanent to his existence. He is, in Lee Rozelle's apt phrase, quite literally "hybridiz[ed]."[40]

This state isn't simply a function of the brutalization of slavery, though the resonances are undeniable. As the Africanist origins of the tale indicate, conjure, too, identifies Henry with the land he works. Does this mean that conjure naturalizes Henry's enslavement? Is conjure merely a tool of the slave trade, as McAdoo's hiring of Aunt Peggy would seem to suggest? The answer is no, even if Chesnutt invites us to say yes. Although conjure appears to regard Henry in the same way that slavery does—as less than fully, discretely human—it actually is antithetical to his enslavement because it insists that no one is "human" in this sense, no one disconnected from or above the natural world. Even Master McAdoo is only one grape away from being, like his supposedly inferior slave, just another part—and casualty—of the vine.[41] Indeed, we all potentially share this fate, at least in the conjure universe, because here we are all enmeshed in networks that exceed our bodies, knowledge, and control. This is the essential difference between Peggy's power and McAdoo's: though superficially analogous in their attempts to "control" nature, hers honors—in fact, derives from—the mutual obligations and limitations that his must violently deny. From this perspective (the perspective of the posthuman tale rather than that of the humanist frame, the perspective of Peggy rather than that of Julius or Chesnutt), Henry's situation is tragic not only because McAdoo fails to acknowledge Henry's personhood but also because he fails to realize the insubstantiality of his own—and, thus, the imposture of his (or anyone else's) mastery. In the conjure universe, equality comes from the ground up, denuding all anthropocentric claims to the naturalness of exploitation, whatever its object and whomever its subject.

Myers is therefore correct to argue that "what gives [McAdoo] his sense of mastery—autonomy, individuality, agency—is his construction of a subjectivity that takes its form in opposition to both slaves and nature, each of which is reduced to 'property' in the Lockean sense," and he is equally right to note that the story undermines this construction through the conjure cosmology, which offers an "alternative view of the relationship between the self and the environment that challenges the dominant culture's mastery over both people and landscape."[42] A problem arises, however, when this cosmology's deconstruction of racism via ecological "interconnected[ness]" is made the basis of a vision of African Americans "exist[ing] in harmony with the landscape."[43] What this reading crucially overlooks is the tale's representation of the real consequences of conjure putting people "in touch with the environment."[44] Though "the practice of conjure" does "mak[e] manifest" a "sense of kinship with the land" that "has its roots in the Earth-based spirituality of the slaves' African ancestors," Myers and Rozelle fail to account for the fact that, in "The Goophered Grapevine," this "kinship" results not in a harmonious picture of "the identities of Southern blacks . . . [being] organically intertwined with the land" but in their disintegration. Becoming one with nature here withers rather than roots self-possession. Henry's "border experience" doesn't "exten[d] [him] psychologically or spiritually"; it buries him.[45]

Myers might contend that this is because of the white salesman's interference or McAdoo's original "exploit[ation]" of man and land, and Rozelle specifically attributes it to the "misus[e] [of the] natural environment" by "gross commerce." Again, though this is incontrovertibly true, we should remember that it is Peggy who is responsible for conjuring the vineyard so that those who steal from it will die, Peggy who "intertwine[s]" Henry with the land to disastrous effect.[46] In other words, conjure postulates the *natural* danger of universal interconnection. Even before the salesman arrives, Henry's fusion with the vineyard (recall that his hair comes to "look des like a bunch er grapes") is manifested not only by his growth in the summer but also by his deterioration in the winter. It's thus a break-even proposition at best, and when we factor in the consideration that the vines' susceptibility to injury has fatal consequences for Henry, we can see that unity with "nature" is not as sublime or fulfilling a prospect as Myers (and, to a lesser extent, Rozelle) paints it to be. To be integrated within the world is, in the conjure cosmology, to be open to harm as well as health, vulnerability as well as—*simultaneously as*—strength.[47] And to attempt to humanize this fact by making it an unproblematic source of holistic self-realization is to risk instrumentalizing conjure into only another means of colonization, simply exchanging McAdoo's form of mastery for another.

It's quite telling in this regard that Henry is not a "full" or "rounded" character: we never hear him speak or are given any insight into how he feels about his incredible metamorphosis; he simply changes, growing or withering as the season dictates. Of course, this silence is no doubt partly a mute commentary on his enforced objectification as a slave, but Peggy, who is not enslaved, is presented in equally flat terms. Alternatively, we could see Henry's lack of legible interiority as a function of Julius's event-centered, largely nondialogic narrative style. But insofar as Julius does provide at least some access into McAdoo's mind (as well as moments of hearing him speak), I think we can attribute Henry's deindividualized nonhumanity—the thinness of his psychological substance—to his conjure-induced assimilation into a posthuman world. This makes sense: the autonomous subject is inevitably a casualty of the erosion of the subject–object divide. As an extension of a grapevine, Henry cannot also be a man, at least as defined by the Enlightenment tradition. Because his "self" is essentially outside (of himself), it is either too little or too much to count. And what's true for Henry holds for "us" as well: conjure erodes rather than grounds not only individual persons but separate, atomic personhood as such.

The same objections would apply to Myers's characterization of *The Conjure Woman* as "call[ing] for the conservation of undeveloped land and wildlife in the South," an interpretation that leads Myers to celebrate Chesnutt's "conservation ethic" and Rozelle to label him an "eco-activist . . . [a] pre-environmentalist."[48] Myers and Rozelle are able to guard this claim against charges of anachronism because turn-of-the-century America witnessed both an unprecedented expansion of industrial mining and logging operations that wrought immense environmental destruction and, correlatively, an intensified interest in preservationist and conservationist activism, embodied in everything from the period's natural history to its literature and politics.[49] This was the age not only of the shift in the epicenter of U.S. logging from the Great Lakes to the South of but also of the establishment of Yellowstone as the first national park (1872) and the founding of John Muir's Sierra Club (1892).[50] Muir is especially relevant in this context because his advocacy in support of national parks was motivated by an underlying belief in the sacred power of "Nature . . . [to] heal and cheer and give strength to body and soul alike."[51] In the face of modern industrialization's dehumanizing effects (often figured as a new kind of slavery), nature is an essential restorative. For Myers, then, Chesnutt's conjure is very much like Muir's environmentalism in that both promise a spiritual connection with Nature in which, if we are in a right relationship, "every pulse beats high, every life cell rejoices, the very rocks seem to thrill with life."[52]

Myers and Rozelle, however, give nothing to support the claim of Chesnutt's "conservation ethic" outside of their readings of the conjure tales. More significantly, insofar as the "conservation ethic" is predicated on a certain self-centered, managerial stance toward the world, it maintains the very subject–object divide that both Myers and Rozelle claim conjure displaces. In this regard, it is not entirely removed from what Outka has called "the white fantasy of a plantation pastoral," which required African Americans to be the willingly receptive objects of the white populace's paternalistic care.[53] Against this presence, however attenuated, of a master self, the conjure stories convey a nonecstatic fusion with the land that leaves neither discrete subjects nor their control in its wake. This means that the scope of the conjure universe's power cannot be limited to its reversal of the Enlightenment's division of self from world; it also incapacitates those post-Enlightenment discourses that understand the self's supposed unification with the universe to be somehow productive, affirming, or gratifying.

This does not mean that conjure and ecocriticism have nothing to say to each other. In fact, the conjure cosmology shares much with the posthumanist ecocriticism of Timothy Morton, which disavows the "positive, sunny," and "far too affirmative" harmonies of "utopian eco-language" in favor of "a new ecological aesthetics: *dark ecology.*"[54] First discarding the romantic ideal of holistic equilibrium, Morton's dark ecology next abandons the concept of Nature altogether, trading its reified stability as something outside of ourselves—and thus amenable to objectification, "possession," and "exploitation"—for dynamic, non-totalizable interdefinitions that are "intrinsically dark, mysterious, and open."[55] And if this echoes the sublime, it nevertheless refuses to repeat the conventional reconsolidation of the subject after its unsettling encounter with otherness, thereby foreclosing what Rozelle describes as the "opportunity to be human in [the] most essential and sacred sense of the word."[56] Instead of such an anthropocentric imperative to serve "some all-too human political, cultural, or racial construct," the posthumanist version of the sublime is, in Outka's formulation, an "eruptive moment of radical uncertainty," a destabilizing movement "in which the identities of self and world become energetically interpenetrative" without being "converted into a conclusive meaning" for an eventually detached spectator.[57] One should not conclude from this that "we" do not exist, only that "we" are not a static, easily isolatable given (in the "mesh" of an "entangled" existence, "nothing exists all by itself, and so nothing is fully 'itself'").[58] More relational than self-identical, "we" not only are *surrounded* by a radically expanded social sphere of "strange strangers" (both animate and inanimate), we *are* such assemblies—and thus are inexorably estranged from our putatively individual selves.[59]

This should sound familiar: it describes Henry's condition almost exactly. He is a man grown unfamiliar to himself owing to his essential openness to the "outside" (figured in the recursive trope of "you are what you eat"). Metamorphosed, he cannot separate himself from his context, and the result is his death; his is a union without a livable unity. And though this may seem exceptional, unnatural, or aberrant, the principle of his transformation is, in the conjure cosmology, universal. Though we may not be vines, we have always been rooted, the produce of strange fertilizations. From this perspective, we are all, to borrow a phrase from Bruce Clarke, "transgression[s] of previous ontological categories."[60] But if we share Henry's pluralized, outside-in constitution, other characters from Chesnutt's conjure tales do so much more dramatically, perhaps none more so than Sandy in "Po' Sandy."

The tale relates the events surrounding a conjure woman (Tenie) who metamorphoses her enslaved husband (Sandy) into a tree so that he won't continue to be loaned to other plantations for months at a time.[61] As Myers notes (and Rozelle subsequently echoes), this transformation repeats the theme, featured throughout Chesnutt's tales, of slaves being "conjured into aspects of the landscape," their "bodies . . . conflated with the land itself—and the crops, trees, and animals that inhabit it."[62] Yet the cost of this conflation, as in "The Goophered Grapevine," is death: while Tenie is loaned away from the plantation (thus reliving Sandy's original complaint), the Sandy-tree is cut down to provide lumber for the new kitchen of the master's wife, leaving Tenie to discover upon her return the gruesome remains of "de stump standin' dere, wid de sap runnin' out'n it, en de limbs layin' scattered roun.'" Even more horrifically, when Tenie, hysterically screaming and desperate to apologize to Sandy for her negligence, soon thereafter tries to stop the milling of the log, she is restrained and tied to a post, and the tree—her husband—is "cut . . . up inter bo'ds en scantlin's right befo' her eyes."[63] Once built, the new kitchen is almost immediately razed—the slaves as well as the master's wife believing it to be haunted—with the wood later being used to build a schoolhouse in which, one morning, a student finds Tenie "layin' on de flo,' stiff, en col,' en dead" ("Po' Sandy," 52).

As numerous critics have contended, "Po' Sandy" transparently allegorizes "the dehumanizing metamorphosis of slavery": its reduction of slaves to thinglike commodities; its violent separation of husbands and wives; its literal and figurative "dismemberment of the body of the slave."[64] Julius himself certainly emphasizes this tragic history, attempting to manipulate Annie into not taking the wood from the now abandoned schoolhouse to build her own new kitchen (John subsequently discovers that Julius wants to use the schoolhouse for his congregation's meetings). But if it's clear that the story has an explicit allegorical dimension, then it's also clear that allegory alone doesn't account for the tale's representation of the power and effects of conjure. In fact, at least in the world of the tale, where conjure is a material reality rather than a rhetorical device, Sandy's transformation into a tree is anything but allegorical; it manifests the conjure-cosmology's actual, radical coalescence of self and environment.[65]

And unlike Muir's rapturous desire to become a tree ("How much I should see, and how delightful it would be! Everything in the mountains would find me and come to me, and everything from the heavens like light"), Sandy's fusion is anything but self-affirming: though it momentarily enables him to escape his enslavement, it also inevitably leads to his demise.[66] Logging is obviously the most immediate cause of the destruction of the Sandy-tree, but ecocritical readings, such as Myers's, that emphasize this point—in contradistinction to what they perceive to be conjure's establishment of an organic "harmony" between African American characters and their environments—fail to account for the fact that even prior to being cut down, Sandy's existence as a tree is far from untroubled.[67] Explaining that "Sandy didn' git erlong widout his trials en tribberlations" while in tree form, Julius gives the example of Sandy, briefly turned back into a man, discovering that "he had a little roun' hole in his arm" after a woodpecker had visited the tree-him earlier in the day ("Po' Sandy," 49). Together with a similar instance, in which a turpentine cut made on the Sandy-tree's bark later leaves "a big skyar on his lef' leg" ("Po' Sandy," 49), this example suggests that Sandy's dismemberment at the sawmill is less a departure from the natural order of the conjure universe than an extension of its immanent logic. Unification with the natural world is innately dangerous in its own right, not requiring human intervention to guarantee disintegration. Perhaps most damning to Myers's argument, however, is the fact that nowhere in the story do the African American characters demonstrate the "regard for landscape" that he attributes to them. Quite the opposite: Julius's successful dissuasion of Annie from using the schoolhouse's wood necessitates that new lumber be used for her project. Thus neither the frame narrative nor the tale—nor, I would argue, Chesnutt—are concerned with "resisting the destruction of the environment."[68] In the tale at least, it is not the environment but the human self that is in danger of being destroyed—subject, as it were, to its own form of environmental catastrophe.

These are not principles limited to "The Goophered Grapevine" or "Po' Sandy." In tale after tale, interlineations of characters and environments result in insensibility ("Hot-Foot Hannibal"), perpetual isolation ("Tobe's Tribulations"), physical deformity ("The Conjurer's Revenge"), or, most commonly, death ("The Gray Wolf's Ha'nt," "Lonesome Ben," "The Marked Tree"). In each, "our" selves are lost to bodies no longer our own. "Lonesome Ben" perhaps best captures this axiom, despite the fact that it does not explicitly rely on conjure to make its point. Having run away from his master's plantation and for months eaten nothing other than the clay of a nearby riverbank, Ben eventually returns "home," starving and lonely, only to discover that he is no longer recognized by either his master or his family.[69] Unable to comprehend this bizarre reception, he looks into the river and sees a stranger's reflection—or rather his own face lightened beyond recognition to the yellowish color of the river's clay. Hearing a frog repeatedly call, "Turnt ter clay!," Ben begins to realize that "he had libbed on clay so long an' and eat so much, dat he must 'a' jes nachl'ly turnt ter clay."[70] Though not induced by an act of conjure, we once again see modeled here a basic law in the conjure cosmology: the mutability of persons upon contact with the putative "outside." The act of bringing the "external" into the "internal" that Ben's ingestion of clay represents thus becomes synecdochic in "Lonesome Ben" for the general bridging of the divide separating the two: as with Henry's metamorphosis into a vine or the *Blockhead*'s fusion with stone, once Ben is "turnt ter clay," there are no absolute grounds for distinguishing between his "inner" self and his "outer" world. There is only the continuity of earth, only the literal, mortal equality of common ground. And for the conjure cosmology, it has always been so, the supernatural transfiguration merely making visible the connection that we normally cannot recognize as the clay of our own becoming.

Inevitably, this transformation–revelation of the substance of Ben's person, in addition to isolating him from his former life, also defamiliarizes what should be his self:

> Now he didn' eben hab his own se'f ter 'sociate wid, fer he felt mo' lak a stranger 'n he did lak Ben. In a day er so mo' he 'mence ter wonder whuther he wuz libbin' er not. He had hearn 'bout folks turnin' ter clay w'en dey wuz dead, an' he 'lowed maybe he wuz dead an' didn' knowed it.[71]

Cognitive and bodily estrangement here converge, reactions not to a dissociative external event but to an alienating encounter with *oneself* as a strange stranger. Far from being a threat that can be named (and overcome) as other, the horror—the difference, the sameness—is within; it *is* the within. Not knowing who or what "he" is independent of a body that he does not acknowledge as his own, and having no one else to confirm his existence, Ben can only surmise that "he"—if there even be a "he"—is (unknowingly) dead. It is an Adamic assumption soon literalized: weakened to the point of exhaustion, Ben lies down and dies, whereupon

> de sun beat down on 'im . . . 'til it baked him as ha'd as a brick. An' den a big win' come erlong an' blowed a tree down, an' it fell on 'im an' smashed 'im all ter pieces, an' groun' 'im ter powder. An' den a big rain come erlong, an' washed 'im in de crick, an' eber sence den de water . . . b'en jes' [the same color] as yer sees it now.[72]

Such an ending leaves no doubt that assimilation into the environment (triggered here by an ostensible assimilation of the environment into one's self) occasions obliteration rather than celebration: "Lonesome Ben" doesn't just kill Ben, it *destroys* him. *After* dying (which would seem conclusive enough), he is "baked" by the sun, "smashed . . . ter pieces" and "groun' . . . ter powder" by a falling tree, and then "washed" away by the river, his coloring of the water the only trace of his existence left (and, of course, even the color is not his own). If we acknowledge the role of the wind in this scene (it causes the tree to fall), then we can see that all four elements participate in the excessiveness-unto-redundancy that is Ben's destruction. Like "The Goophered Grapevine" and "Po' Sandy," "Lonesome Ben" overrides the self–world distinction by overwhelming the self. A literal has-been, Ben's name sounds the temporality of his independent identity.

In stark contrast, therefore, to so-called counter-Enlightenment exaltations of the spiritual and material unification of self and world—exemplified in a "philosoph[ical]" treatise on "transformations" and the "redistribution of matter and motion" that John reads to Annie in "The Gray Wolf's Ha'nt," and which Annie significantly rejects as "nonsense"—"Lonesome Ben" manifests the complex combination of racial trauma and posthumanist cosmology that inspirits all of *The Conjure Woman*.[73] Read this way, Ben's literal sense of self-estrangement ("he didn' eben hab his own se'f"; "he felt mo' lak a stranger 'n he did lak Ben") after discovering "himself" to be turned into clay demonstrates a universal principle of the conjure cosmology:

the self's radical contingency and impermanence born of its inexorable susceptibility to the world. His case may seem exceptional and tragic, a monstrous aberration precipitated by slavery, and, to a certain extent, it is; but it is not, according to the conjure cosmology, unnatural. The medium of his being may have changed, but he was always already an open subject. Neither lording over nor harmonizing with the universe, the privileged human self in "Lonesome Ben" and the rest of Julius's tales is, in a word, eroded. Undoubtedly this condemns the fallacious metaphysical principles of the master–slave binary responsible for Ben's prior enslavement (and partially for his self-alienation as well), but his fate also reveals that the cosmos negates the possibility of ever owning one's self. None of us, in the conjure universe, is his own.

And yet the dispossession of our selves—accomplished by their inexorable possession by other things—need not result from or end in death. Although many of the characters in the tales do in fact die as a result of fusing with their environments, their deaths punctuate a point already made in life: integration into larger networks of becoming attenuates the exclusivity of human-being. In this regard, Julius is actually a more complicated example than the criticism allows. It is certainly true that Julius is a subject in a way that, say, the characters in the tales are not. Insofar as his subjectivity is constituted through the tales he tells, he is a subject *because* the characters in the story are not: he speaks whereas they are spoken; their silence gives him a voice; it is the fiction of their lives that makes his self "real." John, however, implies at various moments that Julius, too, is perhaps something other than a normal human subject. In "Mars Jeems's Nightmare," for instance, John says that "Uncle Julius was a marvelous hand in the management of horses and dogs, with whose mental processes he manifested a greater familiarity than mere use would seem to account for."[74] And in "Tobe's Tribulations," John asserts in similar language that "there clung to [Julius's] mind, like barnacles to the submerged part of a ship, all sorts of extravagant beliefs. The simplest phenomena of life were to him fraught with hidden meaning,—some prophesy of good, some presage of evil."[75]

As critics have rightly noted, these observations are run through with paternalistic racism. Although John elsewhere intimates that Julius's oddities can be explained by the historical circumstances of his enslavement (e.g., "he had been accustomed . . . to look upon himself as the property of another. . . . He had been unable to break off entirely the mental habits of a lifetime"; he presented "the curious psychological spectacle of a mind enslaved long after the shackles had been struck off the limbs of its possessor"), the animalistic language in the preceding examples suggests essential difference, even subhumanity (ideas "clung to his mind, like barnacles"; he shows "a greater familiarity" with animals' minds "than mere use would seem to account for").[76] Distinct from that most human of modes of relating to the world ("use"), Julius's uncanny association with horses and dogs is seemingly a *natural* affinity that places him, with them, in the lower rungs of the Great Chain of Being.[77] "Enslaved" by "habit" and custom rather than taking advantage of the freedom he only passively received ("the shackles had been struck off the limbs of its possessor"), Julius accretes ideas rather than thinks them. Such disparagements, of course, are the fuel that have fired scholars' tireless efforts to underscore the ways in which the stories subversively demonstrate—sometimes subtly, sometimes forcefully—Julius's equality with John.[78]

There is, however, a different way to respond to John's comments. Although they certainly are uttered in a spirit of defining otherness as inferiority, I believe we can hold on to John's insight even as we reject the denigration that he (and arguably Chesnutt) associates with it. If we are less committed, that is, to an agenda of securing Julius's status as an autonomous, economically self-interested subject comparable to John, we might read in these statements the faint apprehension of an alternative, posthumanist cosmology. Note, for instance, what John says about the "source" of Julius's "extravagant beliefs": "Some perhaps were dim reflections of ancestral fetishism."[79] In other words, there might be an African origin both to Julius regarding "the simplest phenomena of life" as if they were "fraught with hidden meaning" and to his intimate proximity to animals. Rather than being attributable solely to the dehumanizing effects of slavery or to the racist prejudices of John, such characteristics might be manifestations of the conjure cosmos's reach into the "real world" of the frame narratives. From this perspective, even Julius's continuing to consider himself "the property of another" long after emancipation might be something other than pathological; such a blurring between personhood and objecthood might signal the fundamental ontological truth of the conjure cosmology: "we" exist only in, and only as, relation (a truth embodied, as we shall see in the next chapter, in the mutual possessions Hurston investigates). And rather than justify Julius's enslavement or disenfranchisement, this diasporic principle rebukes the naturalization of any domination.

We can begin to see this in the very thing that John diagnoses as superstitious ignorance, even evidence of a degrading evolutionary link: Julius's oft-repeated concern with not harming nonhuman beings. In the concluding frame narrative of "Tobe's Tribulations," for instance, Julius explains after telling a tale of unreversed human-to-frog metamorphosis that "dat's de reason w'y . . . I doan lak ter see nobody eat'n frogs' laigs." Similarly, following a story in which a man has been transformed into a mule ("The Conjurer's Revenge"), Julius insists, "I doan lack ter dribe a mule. I's alluz afeared I mought be imposin' on some human creetur; eve'y time I cuts a mule wid a hick'ry, 'pears ter me mos' lackly I's cuttin' some er my own relations, er somebody e'se w'at can't he'p deyse'ves."[80] As in Morton's "upgraded version of animism," which contends that "the ethics of the ecological thought is to regard beings as people even when they aren't," Julius's cautiousness is based on the possibility of animals' hidden humanity.[81] We should beware our brutalizations, he suggests, lest their objects be unrecognized "relations" ("'pears ter me mos' lackly I's cuttin' some er my own relations")—reasoning that, taken to its logical conclusion, means we can never know that the "thing" before "us" isn't one of "us."

According to the diasporic conjure cosmology, however, the concern Julius exhibits need not be limited exclusively to disguised "human creeturs"; other things—from spirits to places to inanimate objects—also can be recognized as our "relations" in the fundamental sense of a shared agency, vitality, and vulnerability that is not reducible to an exclusively human standard[82]—a "relation" or kinship of fellow beings-in-relationship. Indeed, the uncertainty over what counts as "human" cuts both ways, toward the extension of "human" qualities to nonhumans as well as toward the "bestialization" of ourselves; if we can't be sure that "it" is not "us," then neither can "we" be certain that "we" are not "it."[83] This doesn't mean that everything is the same, only that all things are prosthetic compositions of difference. Like Henry,

Sandy, and Ben, "we" cannot determine where the "outside" ends and the "inside" begins, cannot demarcate the line between "who" and "what," because "we" are all jumbles of will and automatism, agency and passivity, resistance and pliability, power and vulnerability. Neither slaves nor free, "we" are defined not by autonomy (often bought at a murderous price) but by entanglement, as "we" emerge in rather than escape out of relation.[84] Constraints and conditions of possibility are here indissociable, if not identical.[85]

Consequently, in the conjure universe, being another's property connotes not an injustice to be overcome but a constitutive enchainment in bonds of "responsibility for the Other," where responsibility is not mastery and Otherness is not necessarily outside of ourselves.[86] Were it otherwise, were responsibility to entail curatorial management, then the ontological privilege (and remove) of the human would be preserved, merely transferred in a kind of shell game from destructive domination to ostensibly beneficent stewardship. Instead of such intensifications of the recursive, potentially unsustainable logic of humanism (we must meet the ever-bigger problems we have created by our past solutions with new, ever-bigger solutions), conjure invites "us" to recognize the fact that "we" have power without total control, agency but not absolute command. In conjure practice, something always escapes one's mastery, even (perhaps especially) when one most vigorously attempts to impose it (there is always a counterspell, another conjurer, an unintended consequence or accident).[87] Julius's ethic is thus based in negative action, or perhaps a kind of active passivity (he doesn't eat or beat certain things), and not in a paternalistic management of his objects of concern, because, as his stories show, one can never assume that objects are inert or that we, at times, are not. "We," too, are things that "can't he'p deyse'ves" from being subjected to the contingency of others uncannily bound to ourselves. This is not to romanticize slavery but to disassemble it from the inside out.

5

"IT MIGHT BE THE DEATH OF YOU"

Hurston's Voodoo Ethnography

PUBLISHED THIRTY-SIX YEARS after the appearance of *The Conjure Woman,* Zora Neale Hurston's *Mules and Men* (1935) contributed to the single most important evolution in genre writing on "other" cultures since the advent of local color fiction: the rapid professionalization of anthropological study.[1] The book certainly advertises itself as a work of this new kind, complete with a prefatory endorsement by Franz Boas (Hurston's undergraduate professor at Barnard and subsequent advisor), a host of tales collected in the field, and appendices devoted to "Negro Songs" and "Formulae of Hoodoo Doctors."[2] And *Mules and Men* seems to deliver what it promises: an insider's glimpse of African American folklore and culture presented, in Hurston's phrase, through "the spy-glass of Anthropology" (*Mules and Men,* 1). The text is replete with the people, places, stories, and histories one would expect in such a work: of her childhood Florida hometown in the book's first half, and of New Orleans voodoo–hoodoo practices in the second. Across both halves, moreover, Hurston repeatedly promotes the material she presents as exclusive, authentic, and representative, an assessment with which many critics historically have agreed.[3]

And yet, despite these professed intentions and the authenticating materials included to fulfill them, much of the recent criticism on *Mules and Men* has centered on the manner in which Hurston actively disrupts the ethnographic genre she ostensibly inhabits. Disregarding her self-avowed status as an ideal, chiasmatic informant–recorder, conversant in the vocabularies of both provincial folklore and metropolitan ethnography, these readers claim that Hurston, like Chesnutt before her, only seems to conform to the generic expectations she announces herself as meeting, all the while actually signifying on her white audience's desire to receive the consumable "truth" of rural African American life. As evidence, they point to Hurston's famous blurring of the line between participation and observation (the book features her as both reporter and character);[4] her related "collaps[e] [of] the boundaries between fact and fiction";[5] and the fact that her collected tales (or "lies," as they are tellingly called throughout) recurrently thematize the confounding of "the white man's surveillance" of his racial others.[6] For these reasons, Hurston is said not only to *report* but to *deploy* an African American cultural strategy of misdirection in which, in Hurston's words, "the Negro, in spite of his open-faced laughter, his seeming acquiescence, is particularly evasive. . . . The Negro offers a feather-bed resistance. That is, we let the probe enter, but it never comes out" (*Mules and Men,* 2). As indicated in the almost imperceptible shift from "the Negro" to "we," Hurston is seen as using the very "tactics" that she attributes to African Americans in general, tactics through which the white public—"always trying to know into somebody else's business"—"can read my writing but they sho' can't read my mind" (*Mules and Men,* 3).

The scholarship is divided, however, as to the intent behind these subversions. Some regard them as a way of "empower[ing] the black self in a cultural milieu where blacks were subordinate" (Amy Fass Emery), even as a means of picturing the "racial health" (Alice Walker) of the generative "black subculture" against the "imaginative wasteland" of whites (Robert Hemenway).[7] Hurston defies ethnographic conventions, in other words, so that she might affirm black identities.[8] Others cite the well-documented inconsistency of Hurston's reflections on blackness to argue that Hurston shows up the inadequacies not only of racism but also of racialism, of any system of thought premised on fixed racial distinctions. Barbara Johnson, for one, contends that, in *Mules and Men* and elsewhere, Hurston "represent[s] difference in order to erase it," thereby underscoring the falsity of early-twentieth-century ethnography's reliance on race.[9]

If these positions are not mutually exclusive, then neither are they easily harmonized. We can say, however, that whether critics argue that Hurston transvalues or deconstructs the racial distinctions that are the occasion of *Mules and Men*'s publication, they share a common conception of the authorial subject as the locus from which the text's antiethnographic strategies originate. Both, that is, regard the intending self—Hurston—as the final hermeneutic though which to assess *Mules and Men*'s deviations from traditional anthropological norms. In fact, many of these critics argue that Hurston distorts ethnographic practice precisely because she wants to articulate a more genuine version of herself than its conventions allow—however that self may be understood.[10] Identity may be performatively mutable or it may be categorically rooted by race, gender, and class, but that it is—that Hurston is—is never in doubt; if the *ethnography* of Hurston's autoethnography is voided, then the *auto-* (its agency, autonomy, and centrality) remains secure. Applauding an ethical praxis of resistance, these readers are necessarily committed to the notion of a coherent, opposition-minded author.

Undoubtedly, much of *Mules and Men* is best understood in this way, as a conscious critique of early-twentieth-century ethnographic practice and its racist–racialist assumptions. Not all of the book's idiosyncrasies can be interpreted as defiant, empowering acts of self-expression, however, and neither does the text always support the complementary paradigm of subjectivity that such readings subtend: a completely realized, wholly sovereign author–agent. *Mules and Men* offers moments that illustrate a very different version of the narrating subject and its range of potentiality—moments that fundamentally unsettle author and text. I thus propose an alternative interpretation of some of the book's unorthodoxies. To do so, I invert Henry Louis Gates Jr.'s paired arguments—that "Hurston's writing depends for its strength on the text, not the context" and that *Mules and Men* can be regarded as a "journey" of its protagonist from "object to subject"—to claim that moments of Hurston's book model a conception of self and text radically delimited by material and spiritual context, indeed, a narrating self so determined by external incidence as to be transformed, contra Gates, from subject to object.[11] Rather than an autonomous subject everywhere offering subterranean criticisms, *Mules and Men* at times suggests a more comprehensive explanation for many of the text's remarkably nonethnographic episodes.

More specifically, I argue that it is the text's insistence on the consequences of inhabiting the voodoo universe that circumscribes (and enables) both the epistemological possibilities of Hurston's anthropological project and the ontological horizons of the self operating within it. Different from *The Conjure Woman* in form, content, and context, *Mules and Men* nevertheless brings to logical conclusion what Chesnutt's tales picture but their frame narratives can't fully countenance: the inseparability of selves, stories, and cosmos. Despite its vibrant actors, boastful characters, and potentially politically minded author, the text retains a voodoo-derived conception of the self's radical contingency, vulnerability, and impermanence. Leaving behind the self-sovereign agent described by Hurston critics, therefore, I explore those moments when Hurston's self is something both more and less than we would expect—and not because she intends it to be—something less authorial because more ethereal: a self fundamentally subject to its physical and spiritual surroundings.

I am advancing, in other words, a reading of *Mules and Men* that suspends the false-surface/meaningful-depth model of significance promulgated by so many Hurston critics—and the correlative performative agent that attends it—in order to give a pointedly superficial reading. In such a reading (inspired in part by Rita Felski's discussion of the phenomenological intersections of "enchantment" and "shock" in *Uses of Literature*), moments of the self's fragility and susceptibility are preeminent, moments when the self not only fails to govern its relation to the world but actually is overwhelmed by it.[12] Neither mere ideological ruse nor aesthetic fantasy, Hurston's hoodoo cosmology means the ineluctable overcoming—the other-coming—of person and text. Voodoo is thus not (only) a "metaphor" or "a figurative concept."[13] Possession, when not "simulated" or "feign[ed]," is not "a blind for self expression."[14] To claim otherwise, to insist that Hurston's engagement with voodoo must be symptomatic of something else, runs the risk of perpetuating the logic of many early- and mid-twentieth-century anthropologists, who claimed, unlike Hurston, that voodoo rites could be reduced to a compensation for historical frustrations or to abnormal physiological or psychological causes.[15] For her part, Hurston was raised among hoodoo and professed respect for—and something approaching belief in—voodoo throughout her life, featuring both throughout her fiction, drama, and memoirs.[16] The relationship was necessarily vexed, however, given Hurston's prizing of "absolute individualism and consequent self-reliance" above all else.[17] An initiate in various hoodoo and voodoo traditions, she finally abandoned them for fear that they threatened her life.

In addition to *Mules and Men,* I also examine this cosmology's articulation in *Tell My Horse* (1938), Hurston's ethnographic record of her time spent studying Obeah in Jamaica and Vodou in Haiti under a 1936–37 Guggenheim Fellowship.[18] Conjure, hoodoo–voodoo, Obeah, and Vodou are by no means identical belief systems, but owing to their common origin in African traditional religions, they share many principles and a similar understanding of the self's place in the universe.[19] Hurston herself, while acknowledging their differences, combines them into one transatlantic network of diasporic spirituality that includes "the Southern United States, the British West Indies, . . . Haiti" and "West African tradition" (*Tell My Horse,* 378, 309)—thereby comprising what Annette Trefzer calls "a global trans-African identity."[20]

I will compare this diasporic cosmology to phenomenologies of spiritual experience from Hurston's time and our own, as well as to Maurice Merleau-Ponty's concept of "flesh" and Heidegger's notion of "care," to reveal its resonances with certain cognate representations of encountering nonhuman alterity. Hurston's voodoo, however, refuses to make otherness an exclusively external condition; here, the self, like a *loa*-dedicated altar or ritual, is an amalgamation of heterogeneous materials and forces that conjure identity as a relationship, a transaction, a profound exchange of powers, possibilities, and obligations.[21] It—"I," "we"—is a syncretic process as much as a possession, the terms of which both exceed and constitute "us," empower and threaten "us." Simultaneously life giving and death bringing, it is an emergent, contingent, ongoing construction that always could be otherwise. As in Poe's and Adams's universes, then, "we" are not our own, but here our dis/possession is not subject to a single, predetermined end (namely, the extinctions of absolute contraction and absolute diffusion, respectively) but rather to an open-ended procession of myriad agencies in fluid relations of command and obeisance, attraction and fear. Cosmological, not monological.

In this way, Hurston's voodoo, like Chesnutt's conjure, dismantles the metaphysical buttresses not only of racism but also of anthropocentricism, of any justification of violent ontological discrimination. Voodoo need not be read as an allegory or political masque to realize this potentiality; it does so according to its own immanent principles. Power inequalities still exist in voodoo, of course, but they cannot be used to authorize the severance of the mutual responsibilities and respect that, the cosmology insists, accompany interconnection. Voodoo may make us instruments of the *loa*, but this is a privilege of gods, not men (and even the gods are bound in service to each other and to "us"). Indeed, the common "flesh" or "clay" that allows Hurston on multiple occasions to feel the pain of nonhuman animal sacrifices *as if they were herself*—as if she was an impossible composition of "them" and "I"—suggests a voodoo principle of mutual becoming that might transcend the violence of some voodoo practices. Departing not only from Poe's and Adams's eschatologies but also from Chesnutt's death-soaked depiction of slave-era conjure, Hurston's voodoo here offers a version of sacrifice embodied not in death but in life, in the enduring, reciprocal obligations and oblations that body forth our being. Such a vision of ontological indebtedness might easily slip into a romantic celebration of holistic self-realization (or even a perverse echo of liberal economic selfhood), yet Hurston's example makes clear that integration doesn't equal consolidation.

A phenomenon in *Mules and Men* well illustrates this dynamic, though it is one that critics have long read as a privileged mode of strategic noncompliance: the narrative's caesurae, gaps, and breaks. In concert with the sustained association of voice and agency in the critical literature on Hurston (as on Chesnutt), those times when her text falls silent—when the narration abruptly ends or when the narrator confesses to being no longer able either to comprehend or to describe her circumstance—frequently have been seen as passively resisting ethnography's insistence on complete access. Read more literally, however, these silences, far from representing a scheme of self-constituting resistance, can be viewed as occasions of the sovereign self's momentary undoing. If expression and selfhood are linked, if the subject writes herself into existence, as Gates contends, then there may be grave consequences when she loses her voice. Even more unsettling, such silences raise the possibility that when we speak, it is not always with our own voice.

DRAWN FROM HURSTON'S 1928–29 field research in New Orleans and published in earlier form in the *Journal of American Folklore*, the

"Hoodoo" section of *Mules and Men,* like the book's introduction, advertises itself as granting unprecedented access to African American culture, specifically here to the rites and rituals of the "secret adherents" (*Mules and Men,* 183) of the hoodoo religion, which previously had been "bound in secrecy" (*Mules and Men,* 185).[22] Dismissing as artificial the bastardizations of "Broadway and popular fiction" (*Mules and Men,* 185), Hurston promises the "reality" of Hoodoo—the significance rather than the signifying of the ritual dances—and admission to "the most inner circle" in which "no layman ever participates, nor has ever been allowed to witness" (*Mules and Men,* 239–40).[23] Hurston, moreover, is careful to stress that "Hoodoo, or Voodoo" (*Mules and Men,* 185), not only comes from "the west coast of Africa" (*Mules and Men,* 227), and thus "keep[s] alive the powers of Africa," but also "is burning with a flame in America, with all the intensity of a suppressed religion" (*Mules and Men,* 185).[24] *Tell My Horse,* too, emphasizes voodoo's link to "the old, old mysticism of the world in African terms" and not to "the catholic calendar of saints done over in black as has been stated by casual observers" (*Tell My Horse,* 376).

As Arnold Rampersad and others have noted, part of the exoticism of this language was meant for the patron of Hurston's early research, Charlotte Mason, widow of "an expert in parapsychology" and an adherent in her own right to a "radical belief in . . . occultism, . . . voodoo and other forms of African religion." Mason, Rampersad contin ues, was "contemptuous of European rationalism and . . . a champion of the notion of the artistic and spiritual superiority of the darker races," and it was this constellation of beliefs that led her to finance Hurston's research for the second part of *Mules and Men.*[25] Although "often express[ing] disdain for white people elbowing in on black culture," she was convinced, according to Hurston biographer Valerie Boyd, "that black people—if they'd only be their 'savage' selves—could save whites from the aridity of civilization. Mason believed in cosmic energies and intuitive powers, and she was sure that 'primitive' people . . . were innately more in tune with these supernatural forces than were whites."[26]

In this belief that diasporic spiritualities (and, by proxy, persons) might revivify an increasingly moribund West (particularly in the wake of the First World War and the rise of Taylorist industrialization),

Mason represented a shift in attitude among certain educated northern whites in the 1930s.[27] Where previously voodoo and conjure were almost universally castigated either as demonic influences or as signs of black peoples' cultural and/or biological inferiority (thereby providing justification for the initial U.S. occupation of Haiti as well as for contemporary anti-immigration laws), they now were celebrated—at least by some—as exotic spiritual and aesthetic salves.[28] Though condemnations and garish horror-show depictions certainly remained, the popular film, music, and journalism of the 1930s increasingly presented voodoo as a positive means of abandoning oneself to a primal ecstasy.[29] Undeniably racist in many of its manifestations—from the notion that African Americans were innately closer to the primitive, sensual energies of nature than their overcivilized white counterparts to the outright exclusion of African Americans (other than musicians) from white-owned Harlem jazz clubs—this newfound interest represented an attempt to engage a completely alien worldview. Yet despite its ostensibly ameliorative and transcultural aims, this cooptation often became the spiritual equivalent to America's seizure of Caribbean and Pacific lands in the first decades of the twentieth century, a synonymous consumption of foreign "resources" for personal and national benefit. Many white devotees of African diasporic spiritualities, that is, reproduced the zeitgeist of self-gratification through Other exploitation, notwithstanding the fact that they often understood themselves to be opposing international imperialism as well as intranational racism.[30]

In some ways, Hurston delivered exactly what Mason, the "godmother" of the Harlem Renaissance, desired: diasporic beliefs, rituals, potions, charms, aesthetics, and more, all presented in such a way as to emphasize the spiritual alternative they offered to the "European rationalism" she so abhorred. Hurston even devotes a chapter of *Tell My Horse* to a "white man who is a . . . Voodoo priest" in Haiti (*Tell My Horse*, 519), describing his soul as "c[oming] from Africa" (*Tell My Horse*, 520) and documenting the times when she witnessed "him in the grip of the African loa (spirits) known as possession: that is, the spirits . . . entered his head and dr[ove] his own consciousness out" (*Tell My Horse*, 525–26). "A new personality burned up the one that had eaten supper with us," Hurston continues. "Before our very eyes, he walked out of his nordic body and changed. . . . Africa was in his tones. He throbbed and glowed. He used English words but he talked from another continent" (*Tell My Horse,* 530). Although Hurston obliquely associates this man, the superintendent of an insane asylum, with the behavior of his charges, Mason would have read this conversion as a vindication of her hope in Africanist self-transcendence, and many reviewers of *Mules and Men* and *Tell My Horse* certainly seized, for praise or censure, on Hurston's ability to effect a similar kind of transport in her readers, "to take us all the way back to Africa—and darkest Africa at that," owing to the fact that voodoo "is in her blood."[31]

In other respects, however, Hurston's presentation radically departs from what Mason and those similarly inclined would have expected from books on African diasporic spiritualities at this time. And the surprise of *Mules and Men* and *Tell My Horse* would not only have been their distance from contemporary ethnographic norms but also, and more dramatically, their distance from the period's typical representations of voodoo. Rather than picturing an exotic conduit for attuning oneself to the oceanic harmonies of the cosmos or cathartically releasing one's repressed desires (the first edition of Freud's *Civilization and Its Discontents* was published in 1930; *Totem and Taboo* in 1913), Hurston offers a universe perpetually disruptive of ourselves. For her, voodoo is not a therapeutic means to consolidate the self by momentarily abandoning it, the encounter with "savage" otherness only a temporary, bracketed occurrence; it is instead a cosmology that permeates all aspects of personal-cum-communal existence—to profoundly destablizing effect.[32] If it defies the rationalism Mason despised, then it also explodes the positive alternative for which she hoped, foreclosing both the idea that human selves are inherently distinct from or superior to their nonhuman environments *and* the seemingly antithetical (but actually coextensive) notion that the self can self-constructively lose itself to the world.

This is not to say that Hurston doesn't give examples of individuals using hoodoo and voodoo for their own gain: in addition to the few cases of charlatanism that she mentions, *Mules and Men* also describes genuine (and apparently effective) rites prescribed for everything from fixing court trials ("law cases" [*Mules and Men,* 223]) to winning lovers ("Man and woman cases" [*Mules and Men,* 187]) to killing enemies or driving them insane ("case[s] of death-to-the enemy" [*Mules and Men,* 239]), and *Tell My Horse* similarly discusses "the power[s] to take human life," "make big cures," "find a job," and be "protect[ed] . . . in order that nothing will happen to you" (*Tell My Horse,* 442).[33] Such individual ends, however, necessarily entail the undermining of individuality as such: to cause persons to do what they don't want to do (whether for "love" [*Mules and Men,* 187] or "justice" [*Mules and Men,* 234]), to infest their bodies with "scorpions" and "gophers" (*Mules and Men,* 186) and dissolve their minds in "hysteria" (*Mules and Men,* 235), to end their lives with a "dance" (*Mules and Men,* 239)—to do all this using only cloth, animals, blood, movements, and incantations is to presume a degree of interconnection with the physical and spiritual universe that destabilizes "our" selves by annihilating our received notions of the self's autonomy, agency, and coherence.[34]

And it's not that our independence is diminished in an ontological vacuum. If "we" are less, it's because other things are more, with agency and vitality now distributed to the nonhuman world; if hoodoo and voodoo rituals are believed to work, it's because their apparently innocuous components, when assembled together, actually possess unexpected, invisible powers over "us."[35] Here spirits can "live in" trees (*Tell My Horse,* 297), obeah sticks ("People whisper its name as if it were a person" [*Tell My Horse,* 321]), "stones" (*Tell My Horse,* 400), and "heads of streams" (*Tell My Horse,* 500), among other "various repositories of the gods" (*Tell My Horse,* 416). Astonished at such pluralism, Hurston asks how a "gourd sheathed in beads and snake vertebrae, and sometimes containing a human bone, came to be . . . the voice of the gods" (*Tell My Horse,* 397). How could a gourd house a god? How could a hollowed shell contain so much?

Elsewhere, however, Hurston shows that we, too, are thus inhabited, "mounted" by *loa* who become "the master[s] of [our] head[s]" (*Tell My Horse,* 389, 448); "do[ing] nothing of [our] own control," we are "the horse of the loa until the spirit departs" (*Tell My Horse,* 495).[36] Consequently, we, like other altars, are motley assemblages or "bag[s] of miscellany," "jumble[s] of small things priceless and worthless."[37] Indeed, in obeah belief, even when we are not obviously possessed, we are still held to be composed of "foreign" materials: "the most powerful part of any man" is not properly of him but an alien agency (a "duppy"), which, however, is paradoxically also the very thing that gave life "to [his] parts" (*Tell My Horse,* 313). Not just *connected* to other things, "we" *are* such constitutive yet estranging relationships, heterogeneous networks nested within like networks. Dispossessed of our own private selves, "we" see ourselves anew as dynamic confluences of bodies, forces, and events, none of them wholly our own, none reducible to "us"—even as the us–not us distinction becomes so shadowy as to preclude definitive demarcations (in the dance of possession, for example, "the drums and the movements of the dancers draw so close together that the drums become people and the people become drums" [*Tell My Horse,* 323]). What voodoo reveals, therefore, is the impossibility of escaping the influence of one's environment to master it, of existing at an imperious remove from one's milieu, because here the environment is no longer an external reality opposed to ourselves but instead the very material of which "we" both are and, then, are not. Caught in a cosmos that makes profound even the most incidental connection, bound in relationships where everything potentially has a soul that might impinge on—even displace—our own, "we" are as much manipulated as manipulating, as much acted on as acting. A two-way possession (and perhaps more).[38]

In Hurston's account of voodoo, then, power is borrowed at cost, our selves—bric-a-brac that they are—mortgaged. When we approach the gods with a request that they serve us ("I disturb you to ask you to serve"), they demand payment in kind ("I know what you want, but are you disposed to serve me also?" [*Tell My Horse,* 461]). "The Petros work for you only if you make a promise of service to them" (*Tell My Horse,* 441). To claim power, in other words, we must be "claimed" in turn ("the spirit . . . has claimed the candidate" [*Tell My Horse,* 448]; candidates must "permit the spirit to dominate them" [*Tell My Horse,* 451]). To become "invulnerable" in one way depends on being vulnerable in/to others (*Tell My Horse,* 448). There is thus no isolated or stolen agency, no strength without weakness, only mutual service, meaning that even a protection spell isn't an escape from relation but an alternate form of it. Consequently, voodoo rites delineate the limits, not the limitlessness, of our power, as their efficacy depends on the often dangerous process of sacrificing our persons to forces greater than—yet intimately indissociable from—ourselves.

Even Marie Leveau, greatest of the New Orleans voodoo priestesses, is said to have been able to access the power needed to answer people's requests only by, in her student's words, "go[ing] to her great Altar and seek[ing] until she become the same as the spirit.... She answer them as a god.... Marie Leveau is not a woman when she answer the one who ask. No. She is a god, yes" (*Mules and Men*, 195). The crucial thing to note here is that, although Leveau assumes the power of "a god," she is only able to do so by losing her separate self, a nonautotelic process that logically culminates in Marie prophesying, through this same power, that she will soon die (*Mules and Men*, 194).[39] It is a paradoxical power, prophesying one's own death, yet perfectly emblematic of the logic of voodoo's spiritual economy: accessing the spiritual energies with which one is interlineated inevitably ends in the loss of one's individuality, interiority, and, eventually, life. Voodoo is a "power more than equal to man" (*Mules and Men*, 225).[40]

Hurston herself soon experiences the mortal dangers of delving into the hoodoo universe. And it's not as if she didn't know any better: throughout the second halves of *Mules and Men* and *Tell My Horse*, numerous people repeatedly warn her that she risks her life by investigating voodoo, with one friend bluntly cautioning, "Ah wouldn't mess with it if Ah wuz you.... It'll kill you" (*Mules and Men*, 187). Similarly, a hoodoo doctor, prior to initiating Hurston into the mysteries of his practice, counsels, "It might be the death of you" (*Mules and Men*, 239), and Haitian acquaintances suggest that her voodoo research might get her "involved in something... terrible, something from which [she] could not extricate [herself] alive" without losing her "humanity" (*Tell My Horse*, 470, 481). At one point, Hurston's Haitian maid even blocks her from leaving her lodgings at night to follow the sound of "drums throbbing at a distance" (*Tell My Horse*, 475), explaining, "Some things are very dangerous to see" (*Tell My Horse*, 476).

Although Hurston does not, in fact, die from her contact with hoodoo, she nevertheless loses her discrete self to it—even if only momentarily. When Hurston experiences the ritual of "getting the Black Cat Bone"—"most terrible of experiences" (*Mules and Men*, 220)—she enters a state of being that leaves her incapable of either clearly narrating or understanding the events unfolding around her.[41]

After being warned to "look out! This is liable to kill you," she can only recall that

> [her initiators] communicated some unearthly terror to me. Maybe I went off in a trance. Great beast-like creatures thundered up to the circle from all sides. Indescribable noises, sights, feelings. Death was at hand! Seemed unavoidable! I don't know. Many times I have thought and felt, but I always have to say the same thing. I don't know. I don't know.[42] (*Mules and Men*, 221)

Here the description of the ritual ends, with no further details or analysis provided. According to Boyd, this reticence remained constant throughout Hurston's life: she "always recounted the Black Cat Bone experience with excessive . . . vagueness."[43]

What are the implications of this caesura, this violent disruption of the text's own narration of violence? On one hand, Alice Walker and others have read it as authenticating Hurston's ethnographic work by demonstrating the unique extent of her cultural embeddedness: even if it breaks down, Hurston's narration is of a ritual never before described in the first person. Yet because this occasion and others like it instantiate the interruption, not the realization, of ethnographic knowledge, it seems paradoxical to value their resultant silences, loud as they are, as epistemological advances. On the other hand, such aposiopetic moments have also been read, by Wendy Dutton, among others, as willful subversions of ethnography's aspiration to an encyclopedic knowledge of other cultures.[44] Extending Hurston's description of "the profound silence of the [voodoo] initiated" to Hurston herself (*Mules and Men*, 185), Dutton interprets the mystery of the scene, its tantalizing silence, as a politically charged denial of information to Hurston's white readership. As in the case of Chesnutt and Julius, Hurston's silence is here read as a positive signifying practice, a withholding, a fitting complement to the tales' narration of more explicit tactics of evasion and defiance.

I agree that this moment challenges ethnographic theory and practice. But what if we were less certain of Hurston's control over it? What if we entertained the possibility that the challenge (and the fear and confusion that compose it) originates in something other than strategy or ideology? What if Hurston is actually telling the truth, honestly reporting as best she can an event that overwhelms her? What if she *believes*? What if she really did go "off in a trance," really did fear for her life in a moment of "deadly peril" (*Dust Tracks*, 700)? What if intelligible reporting breaks down here because the detached, sovereign author-function breaks along with it, among a profusion of incommunicative "I don't know[s]," their empty, chantlike repetition sounding the hollowness of the "I" that utters them?

As in an earlier episode that concludes the first half of *Mules and Men* ("Folk Tales"), where Hurston is forced to hurriedly leave her Florida tale collecting because a woman attempts to stab her at a party, the comprehensive ethnographic project that should guarantee full disclosure is here disrupted by Hurston's inability to abstract herself from the events in which she participates.[45] Once more, Hurston's vulnerability necessitates a cessation, a caesura—her version of Henry Adams's self-spectralization. Unlike in the Florida incident or in Adams's entropic science of history, however, here the threat immanent in Hurston's environment is invested with all of the awesome portent of the voodoo metaphysics she bodily explores (the snake skins she wears, the cat bones she tastes, the blood potions she drinks).[46] As in Paul Outka's and Sharon Davie's respective analyses of *Their Eyes Were Watching God*, a "representational lacuna" or "sudden, eruptive, failure of language" here results from an *impersonal* "sublime experience . . . in which everything is temporarily called into question"; it is an experience "which one cannot pin down, fix, with rational thought, . . . which one can name only to unname . . . or to fail at naming, or to succeed through admitting failure."[47] Consequently, what is at stake in this instance of bio/graphical dissolution is not merely "the death of the ethnographer's" subject-position—which might eventuate in a more authentic and imaginative relation to one's self, as John Dorst has argued—but also the demise of the coherent, self-determined subject altogether, a potentiality embodied by the very real possibility of Hurston being killed by voodoo: "Death was at hand! Seemed unavoidable!"[48] As Hurston's biographer Robert Hemenway notes, she feared such a fate outside of the text as well, prematurely ending a later exploratory trip to Haiti because she was afraid that the subject of her research was threatening her life: "Hurston was convinced that [a sudden, serious] illness and her voodoo studies were related."[49]

The threat of voodoo killing Hurston poses immense difficulties not only to Mason's self-serving instrumentalization of diasporic spiritualities but also to philosophical studies of ecstatic religious experience in the 1920s and 1930s. Mostly focused on Christian (or at least monotheistic) mystical traditions, these accounts followed William James's *The Varieties of Religious Experience* (1902) in depicting the general "healthy-minded" "optimism" of ecstatic union (they most often dropped James's hard-won *achievement* of such faith, however).[50] James Pratt's *The Religious Consciousness* (1920), for instance, describes "the emotion of the ecstasy" as being above "all else a burning love for God and a joyous consciousness of this loving presence." "Com[ing] into contact with a reality other than [one's] own consciousness" may momentarily lead to "sudden fear" or even the "interrupt[ion] . . . of one's thought" and the "derange[ment]" of one's sensations, according to Pratt and others, but this inevitably resolves in the boundless peace, "joy[,] and love" achieved in perfect communion with God.[51] Mysticism, to be considered successful, must end with "the individuality of the mystic becom[ing] merged," sublimely, with "the Universal Self."[52] As Rudolf Otto famously proclaims in *The Idea of the Holy* (1923), any religion that does not attain this transcendence, that does not move beyond the "fear" and "dread" of first encountering the numinous "*mysterium tremendum*" or "Wholly Other," marks itself as a "crude, barbaric antecedent" to more advanced faiths.[53] In other words, it is good and proper that we are struck dumb before the divine, that "the *feeling* remains [when] the *concept* fails," but it's an unnatural return to the "primitive"—and its "wild and demonic forms"—if this feeling is dominated by "the 'fearful' and horrible."[54] To not progress through well-defined stages and arrive at a concrete, pacific telos, to fail, relatedly, to divide the "sacred" and the "profane" into discrete spatiotemporal categories, is to be judged culturally (albeit sometimes therapeutically) degenerate.[55]

If this dismissal of "primitive" beliefs has largely faded in more recent historical and phenomenological studies of religious ecstasy, the assumption that such experiences move from "violent" beginnings to a paradoxically self-affirming "undifferentiated unity" remains prevalent.[56] Perhaps surprisingly, given the different religious traditions, this arc also characterizes the host of recent interpretations that read *Mules and Men* and/or *Tell My Horse* as a "recuperative project," as "an initiatory quest" in which Hurston finds through voodoo "an occult spiritual authority" and a "mythic agency" that grant her both "knowledge and power," or simply as the story of Hurston "finding . . . herself in the black world she described."[57] More generally, discussions of voodoo from the past few decades also typically have presented it as "affirming personhood," facilitating a "process of self-discovery," or providing a means for "individuals [to] actualiz[e] themselves." According to these evaluations, voodoo does not "destr[oy] the integrity of the self . . . [but instead] provides an increased scope for [the self's] fulfillment."[58]

These accounts aren't necessarily wrong, only incomplete. They rightly note, with Hurston, that "Voodoo is a religion of creation and life" (*Tell My Horse,* 376), but they fail to acknowledge its accompanying dangers. They don't grapple with the fact that she is threatened rather than confirmed by her contact with voodoo. As Hurston herself repeatedly insists, "voodoo . . . in Haiti . . . [is] both beautiful and terrifying" (*Dust Tracks,* 711), "triumphant" and "frightful" (*Tell My Horse,* 411). It combines "joy and ecstasy" (*Tell My Horse,* 434) with "terror . . . so humid you could smell it and feel it on your tongue" (*Tell My Horse,* 412).[59] This explains why adherents who approach *loa* "wish[ing] to gain something . . . fear them at the same time" (*Tell My Horse,* 441). Like the clay in Chesnutt's "Lonesome Ben" or the Jamaican Marjo Bitter vine Hurston describes in *Tell My Horse,* the voodoo gods are *pharmaka*: by one measure "a most excellent medicine," by another a "violen[t] poiso[n]" (*Tell My Horse,* 298). They encompass life and death, other and self, a dynamism—equally ontological and affective—that accords with Hurston's personal conviction that "destruction and construction are but two faces of Dame Nature" (*Dust Tracks,* 795). And the Black Cat Bone scene captures both, as Hurston is at the same moment emptied and overfilled, vitalized and vitiated. Against the emphatically self-gratifying conception that voodoo–hoodoo must empower us without risk or payment, therefore, *Mules and Men* offers the stubborn evidence of itself; when it stops speaking—or, in what amounts to the same thing, only describes "indescribable noises, sights, feelings" or repeats "I don't know"—it is because the voodoo universe has made its author no longer fully available to dictate her life or her text.[60] A Hurston still exists, of course—after all, someone represents the scene to us, such as it is—but it is a Hurston riddled with as much uncertainty as the representation itself. And here the uncertainty and instability are not momentary, transitional states but the constant attendants of one's being. *Ek-stasis,* not ecstasy.

I would thus argue that the breakdown in rational reporting ("Maybe I went off in a trance"), the series of "I don't know[s]," and the silence that subsequently follows are all symptomatic not of a covert strategy but of Hurston's suddenly indefinite ontological state; the threat, the violence, the fear: these are all dramatizations of a person confronting a universe both larger and closer than she imagined. When what defines the self is physical and epistemological mastery over one's environment (or at least distance from it), then not controlling or not knowing means not being. Far from "finding" herself, "actualizing" herself, or "fulfill[ing]" herself, therefore, Hurston at this moment loses herself to voodoo, the threat of physical death only the belated index of a metaphysical principle already accomplished: even though Hurston survives the scene, the encounter reveals that her heroically individualistic self—discrete, coherent, self-determining—was never alive outside of her fictions.[61]

This is obviously not a religion suited to Mason's self-interested purposes, and possibly not to Hurston's or ours. To say otherwise, to assume that the voodoo–hoodoo universe is merely a technology for finding oneself, is to echo Mason's paradoxical, imperializing desire for something other than herself with which to satisfy herself. And the reference to imperialism is apt because, as with conjure's disabling of the antebellum and Reconstruction racisms depicted in Chesnutt's tales, the hoodoo–voodoo of *Mules and Men* and *Tell My Horse* denaturalizes the white superiority complex behind both America's renewed colonial ambitions and the reascendancy of such groups as the Ku Klux Klan from the 1920s through the 1940s. Although Hurston's ambivalence toward the U.S. occupation of Haiti is well documented (having arrived just after its cessation, she writes that the United States had been "the deliverer of Haiti" and that "nobody wanted the Marines to go away" [*Tell My Horse,* 337, 352]), the diasporic cosmology that she documents powerfully invalidates the philosophical underpinnings of imperialism.[62] It does so not as an ideological allegory, nor solely by the mere fact of its existence (thereby disproving the long-standing claim that people of African descent lacked their own viable culture, history, and religion), but also by its immanent principles: voodoo does not abide the absolute differentiation of servantship and mastery because it defines them as fluid actions, not reified beings.[63] Against, then, the charges of contemporaries like Alain Locke and Richard Wright, who claimed that Hurston's work pandered to white expectations by offering stereotypes of African American folk life without activist commentary, *Mules and Men* and *Tell My Horse* radically incapacitate the metaphysical justifications of racial inequality.[64] It's simply not a form of resistance that Locke and Wright would have recognized.

Also as in Chesnutt's tales, however, the scope of the diasporic spirituality's force in *Mules and Men* and *Tell My Horse* exceeds its dismantling of the white–nonwhite binary; the more foundational distinctions of human–nonhuman, self–other, and subject–object—for which racist differentiations are the most infamous synecdoches—are troubled as well, meaning that "we" cannot assume ourselves, much less our mastery over a supposedly inert, lifeless world.[65] Voodoo delegitimizes the human exploitation of the nonhuman, first by hopelessly entangling the two, and second by positing a universe in which any thing, any place, might be the dwelling of a god. Against the assured ontological priority of "modern man," the so-called premodern sees all of "nature [as] capable of revealing itself as cosmic sacrality. The cosmos in its entirety can become a hierophany [a manifestation of a sacred reality]."[66] Rejecting "modernist metaphysical nothingness" and Christian–romantic metaphysical oneness, voodoo offers an almost infinite plurality of metaphysical registers and agencies.[67] Being—or, rather, becoming—is here defined by abundance, not scarcity. Consequently, any encounter could be a sacred one.

This may seem inconsistent with the actual voodoo ceremonies that Hurston records; forgoing the romantic idea that "the darker races are visiting angels" (*Dust Tracks*, 793) without the human instinct for violence and domination, she repeatedly depicts graphic instances of the "enormous and unconscious cruelty" (*Tell My Horse*, 347) of the

Haitian people toward animals: "the tongue of [a] red rooster torn out before it is killed" (*Tell My Horse,* 434); the "cut[ting] off [of] the testicles" of a live, "groaning" pig (*Tell My Horse,* 444); the "pull[ing] [of] the teeth of [a] dog" before "he was buried alive" (*Tell My Horse,* 447); and "the sickening sound of cracking bones" as live chickens are prepared for the gods, a scene that causes Hurston to respond, "My heart flinched and my flesh drew up like tripe" (*Tell My Horse,* 424). How are we to reconcile such violence—even torture—with the belief in the distribution of deities throughout everyday places, things, and beings? The literal answer is that, according to "voodoo rites," "all gods and goddesses must be fed" (*Tell My Horse,* 388), and therefore some beings—*because they are valued but are not gods*—must be sacrificed as propitiations. But the question remains: how can one tell the difference? Given that Hurston documents cases of animals being treated as divine receptacles, how can one be *sure* that the thing being torn to pieces is not a god in its own right, a being whose demands must be recognized and honored rather than a "miserable creature . . . too weak to hold what we want" (*Dust Tracks,* 766)?[68] Though perhaps unanswerable in any concrete way, these questions gesture to the possibility that the voodoo-inspired recognition of the *potential* sacredness of all things could be pushed beyond actual voodoo practices, made even more universal than the imperatives of blood sacrifice allow. By an immanent principle, voodoo's dense network of obligations, service, and respect could be extended to those animals that otherwise are acknowledged only in death.

Indeed, we need not even insist on the religiosity of this recognition.[69] We could imagine a desacralized version of relating to the nonhuman agencies circulating within, around, and beyond "us"—not only animals but vital places and things as well.[70] Following Hurston, who consistently lamented the widespread belief in African Americans' "reversion to type" (the notion that black people have only "a superficial layer of western culture" that, when scratched, will reveal "the jungle drums throb[bing] in [their] veins"), we could attribute such forces to "Dame Nature" rather than to spirits, to the astoundingly complex but not necessarily supernatural processes of the cosmos.[71] And one possible form that this nonholy but not-necessarily-secular cosmology might take is indicated in Hurston's complex expression that her "flesh drew up like tripe" in reaction to the torture of the chicken and its subsequent "crash[ing] into" her as "it leaped in its death agony" on "broken thighs" (*Tell My Horse*, 424). Specifically, the language calls to mind Maurice Merleau-Ponty's late phenomenology of "flesh," the principle of chiasmatic continuity between self and not-self brought into being by their mutual embodiment: "The body unites us directly with things through its own ontogenesis, by welding to one another the two outlines of which it is made, its two laps: the sensible mass it is and the mass of the sensible wherein it is born by segregation and upon which, as seer, it remains open."[72]

Though this unity of flesh is, for Merleau-Ponty, not based in "matter" (being instead "a *general thing*... a sort of incarnate principle that brings a style of being wherever there is a fragment of being"), and though it does not lead to a final identity of "the seer and the visible" but only to a "strange adhesion" of the two (pictured by a "reciprocal insertion and intertwining of one in the other"), it nevertheless can provide a basis for thinking about the implications of sharing what Hurston calls a "common clay" with the world (*Tell My Horse*, 393).[73] If nothing else, it offers an explanation for why Hurston *literally* feels like butchered meat at the moment a bird's body is broken, the sensation of its fractured fleshliness indistinguishable from her own. And it's possible that such reciprocal touching and being touched might trouble the ease with which "we" justify domination through projected-yet-naturalized differences. If "we" are a site of contact rather than a discrete isolate, conjured by combinatory rituals rather than our own will, then it's hard for "us" to hold on to a sense that otherness is exclusively the property of the other; when "we" feel the pain of the "other"—indeed, when "we" merely acknowledge the liveliness of "other" bodies (even inorganic ones)—"we" experience the twinned, uncanny sensations of their bodies as our own and our own as someone (or something) else's. "Flesh," then, affords a concept of mutual becoming and unbecoming that potentially opposes what Hurston calls "the principle of slavery," which transcends "the literal buying and selling of human flesh" to include "the efforts of any... community to live and advance their interests at the expense of the lives and interests of others" (*Dust Tracks*, 767).

Hurston has in mind here the ongoing fact of "human bondage" despite the West's hypocritical trumpeting of "the dignity of man," but her critique of the perpetuation of slavery under the name of "acquiring sources of raw material [to] kee[p] the market[s] open" (*Dust Tracks,* 790) easily could include "the lives and interests" of nonhuman "others" as well. After all, Hurston mocks the idea that "unspeakable tyranny" is limited only to "enslav[ing] one's own kind" (*Dust Tracks,* 790), and her reason for rejecting "Race Pride" (i.e., why "prid[e] yourself on your physical make up, something over which you have no control"? [*Dust Tracks,* 784]) applies equally well to human constitutions, especially given the strange matter of their making.

From this vantage, we might even see anew the "fiendishness" of zombification (*Tell My Horse,* 456), to which Hurston devotes an entire chapter in *Tell My Horse.*[74] What's "horrible" about the possibility of becoming a zombie (*Tell My Horse,* 456), according to Hurston, is its literal brutalization of the victim, who is changed from "an intelligent being" into "an unthinking, unknowing beast": "working like a beast, unclothed like a beast, and like a brute crouching in some foul den in the few hours allowed for food and rest" (*Tell My Horse,* 457). Yet, as discussed in chapter 3 in relation to Frederick Douglass's conversion from ox to man, might we not reexamine the use of the "beast" as the natural mark of passivity, stupidity, and servitude? While acknowledging the continuing horror of human enslavement, shouldn't we also—by the same means, for the same reasons—consider the violence entailed in subjugating the nonhuman (including the inanimate) world, especially when the commonality of "flesh" exposes the fraudulent hierarchization of "Man"? Might we not abolish the instrumentalizing sign of the "beast" (the natural, to-be-dominated Other) altogether?

Doing so need not entail the humanization of the world (i.e., affording it rights, liberties, etc.) or the erasure of difference (which all too easily could result in—even perversely condone—the murderous unity of flesh achieved in the bloody cycles of Haitian revolution that left "dismembered parts of bodies to which no one could give a name" [*Tell My Horse,* 335], "human fragments," and "hunks of human flesh scream[ing] [in] outrage" [*Tell My Horse,* 336, 334]). The point isn't this genocidal reduction to "formless matter" (*Tell My Horse,* 374) but a recognition of the common flesh behind all forms, so that the "mingl[ing]" of "bodies and parts of bodies" becomes a principle of life and not only of death (*Tell My Horse*, 373). Far from a fatal homogeneity, a voodoo version of flesh offers, in the spirit of Derridean "monstrosity," the prospect of endlessly proliferating difference such that sameness and otherness are discarded in favor of much stranger—because more multiple, ever shifting, and uncertain—borders ("to be a monster," according to Deleuze, "is first of all to be a composite").[75] It goes without saying at this point that these borders include our own. Previously perched defensively above our sovereign kingdom, "we" are now of the masses. Our moats have been filled, our ramparts abandoned, our lordship abdicated.

A vision like this might easily swing from genocide to utopia, to the kind of self-gratifying conceptions of David Abram and others who convert the unity of "flesh" into a medium for experiencing only the "enlivening pleasure of . . . contact" with "the more-than-human cosmos."[76] Again, these accounts rightly capture one dimension ("the joy and even ecstasy" [*Tell My Horse*, 434]) of being inextricably bound to our lively, vibrant environments, but they fail to appreciate, at least according to the voodoo cosmology, the associated costs of such union ("terrible sometimes" [*Tell My Horse*, 430])—thereby promoting, like Mason, a paradoxically *solipsistic* satisfaction of "becoming earth" (Abram characterizes his latest work as "a book of solitudes").[77] The voodoo cosmos is not conducive to cultivating oneself through pleasurably glancing encounters with otherness; it is not a pacific idyll but a diverse field of contests and alliances, demands and accommodations, among entangled actors ceaselessly interacting on the "fluid boundary between domination and resistance."[78] Like Hurston's description of language, this universe "may be full of beauty and pleasure," but it is also "full of death" (*Dust Tracks*, 793)—not in opposition but in combination, via what Edward Pavlić calls "disruption-*as*-renewal."[79]

What's more, the voodoo cosmology does not promise that we will ever be rid of such ambiguity. Unlike in Poe's and Adams's universes, there is no final, conclusive end, no oddly anthropocentric certainty of a foreknown apocalypse, only the hopeful–fearful prospect that things will go on even if "we"—the old, exclusive, tyrannical "we"—don't. According to Dana Rush and others, this is perhaps the defining feature of West African Vodun and cognate diasporic spiritualities: a perpetual "dialectic" of "the ephemeral (impermanent, fleeting, short-lived) and the unfinished (ongoing, enduring, never-ending)." Here, processes and their emergent products perpetually transform each other in a "protean . . . system that thrives on . . . flux and possibility."[80] Consequently, there must always be a remainder to voodoo practice, always a possibility for a reversal of fortune: one cannot know all of the powers that compose our existence (Hurston notes that "new local demigod[s]" are identified all the time [*Tell My Horse,* 377]), and inevitably, one of these unaccounted-for variables will intervene to establish a new, momentary order before it is replaced, in turn, by another (like Chesnutt before her, Hurston recounts numerous situations in which conjure practitioners work to cancel each other's hoodoo). There can be no final conjure, no completed mastery, only the continued *performance* of becoming. "We" never arrive but are always in the sending.

And if the voodoo cosmos is thus "unfinished," then so, too, must be the voodoo cosmology, making both "works in progress, eternally under construction."[81] The lack of closure is conceptual as well as temporal: voodoo is typified by "local variation[s]" rather than central dogmas (Hurston admits that *Tell My Horse* "does not pretend to give a full account of either voodoo or voodoo gods" because "*no one* knows the name of every loa" [*Tell My Horse,* 377, 397, 377]), the corollary of which is that voodoo rites privilege "context and circumstance as opposed to fixed essences."[82] Indeed, every ritual "is always new" because the encounter with the gods "varies . . . each time" (*Tell My Horse,* 411), requiring constant adaptation and renegotiation. This doesn't mean that anything goes in voodoo; rather, its variations are departures from recurring themes, like improvisational riffs off of a bass line. They follow what Hurston calls the diasporic "rhythm of segments" rather than symmetrical "whole[s]," a rhythm that weds creativity and constraint, freedom and compulsion.[83] Nevertheless, no single truth, meaning, or master narrative is adequate to voodoo's plural forms.

How, then, are we to respond? This is an impossible question given the underdetermination of both the voodoo universe and who (or what) constitutes "we," but one possibility is Pavlić's recommendation of "humility and caution," even a little "healthy fear," as a way of checking our attempts at either enslaving or becoming the universe.[84] More generally, we might entertain a posthumanist version of *Sorge*, the "care" that Heidegger says is the essence of *Dasein*'s being-in-the-world and thus the foundation for all individual cares, concerns, and relations—including the "essentially ambivalent . . . condition intermediate between being repelled and being attracted" that Dutch Heideggerian phenomenologist Gerardus van der Leeuw nominated in 1933 as "the primeval experience in religion."[85]

Yet, where Heidegger makes such care-fullness the exclusive privilege of human beings and claims that it is "authentic" only when actively defined against *Dasein*'s "subservience to . . . others" (when *Dasein* is cared for rather than caring for, it is "dominated," its "freedom" and being are "taken . . . away from it"), we can imagine a form that neither presumes a curatorial relationship to the world (i.e., taking care over) nor regards "belong[ing] to . . . others"—whether human or nonhuman—as a "dictatorship" to be overcome.[86] In this voodoo-inflected version of *Sorge*, the description "'I' 'am' not in the sense of my own self, but I am the others in the mode of the they" is not merely an "*initial*," "*primordial*," or "*deficient*" mode of being, a public "disguise" that must be "br[oken] up" and discarded to make the world rightfully one's own, but an inescapable constituent of becoming to which "we"—ontologically schizophrenic—are forever responsible.[87] It may even be that such a plurality hints at an ethics of mutual possession and service, as when Hurston writes in the conclusion of her autobiography, "I have served and been served" (*Dust Tracks*, 769)—to which we might add the reminder that the "I" is itself a kind of service: an incantation before, an enlistment of, a giving over to "foreign" powers. From this perspective, the song that Hurston records at a Jamaican wedding, "Let the Stranger In" (*Tell My Horse*, 285), sounds both an imperative and an always already accomplished fact, one as risky as it is inescapable.[88]

HURSTON ULTIMATELY ABANDONED her voodoo research to save her life, and *Mules and Men* follows suit, pulling back from its unsettling apprehension of the hoodoo cosmos to restore the text to narrative and Hurston to herself: immediately following the portentous string of "I don't know[s]"—each pregnant with the possibility of a nonexistence that cannot quite be borne—the closing line of the chapter domesticates the entire Black Cat Bone experience with an offhanded, even blithe reassertion of Hurston's self: "Before day I was home, with a small white bone for me to carry" (*Mules and Men,* 221). Radically disorienting in its tonal disjunctiveness, this shift from ethereal spirit possession (i.e., being possessed) to banal material possession (possessing a home and a bone and one's self) erases the process of Hurston's recovery from the ritual that her autobiography claims "took [her] months to doubt . . . afterward" (*Dust Tracks,* 700).[89] It also bespeaks the degree of disengagement required to articulate even this much of Hurston's contact with voodoo. That *Mules and Men* exists at all testifies to Hurston's withdrawal from a universe that is incompatible with what she hails as the "gift of individualism" (*Dust Tracks,* 782) or "the expansion of seeking into individual capabilities and depths" (*Dust Tracks,* 786). As with Chesnutt's frame narratives, then, the book marks a tenuous circumscription—simultaneously an engagement and a retreat—around an impossible, ineffable loss of autonomous selfhood (and yet lost by whom or what?). Ontologically rather than (or at least in addition to) ideologically defying all attempts at totalizing knowledge, Hurston's text is the record of an encounter with a universe that can be neither inhabited by our sovereign selves nor escaped with our separate lives. If slave autobiographies cannot long countenance the threat to their self-enunciative projects that diasporic spiritualities represent, then even in *Mules and Men* there is a fundamental tension between the voodoo universe and the attempt to objectively represent it, because to be fully integrated into this universe is to preclude the stability of both author and text.

The drama of *Mules and Men,* therefore, is that it actually threatens to be overwhelmed by its subject. Although the Black Cat Bone scene is an aberration (despite the insistence throughout on Hurston's embeddedness, rarely is it brought to such a crisis), the text wears the scars of its encounters with the voodoo cosmos even, paradoxically, when pulling back from it. When, for instance, the narrative portion of *Mules and Men* concludes with Hurston escaping into the "lie" of a mythical self—"I'm sitting here like Sis Cat, washing my face and usin' my manners" (*Mules and Men,* 246)—it might appear that Hurston has consolidated herself through an empowering identification with the tradition she has been documenting. And yet, if this is so, then it is nevertheless an identification that bears the trace of its—and thus her—disruption. Having as part of the Black Cat ritual "passed" "the bones of the [boiled-alive] cat . . . through [her] mouth" (*Mules and Men,* 221), Hurston's subsequent reference to "Sis Cat" necessarily conjures the violence of the earlier cat's "scream[s]," the "terrified" cries that soon became Hurston's own (*Mules and Men,* 221). As suggested in the earlier discussion of "flesh," this transposition of affect—shared terror and screams—signals an uncanny familiarity-unto-familiality that the voodoo universe literalizes: "we" are not separate from a lifeless world but fearfully, bloodily interlineated with myriad agencies no longer divisible into self and alien, human and nonhuman. As in Poe's "The Black Cat," the border separating "us" from "them" dissolves in the moment when our fates become, in the very moment that the humanist "we" un-becomes, one. Hurston's discrete self thus must always be a myth.

In these turbulent moments in which it does not fully exist, when it falters to hold on to its sense of self by repeating a nonphrase ("I don't know") that always already threatens its existence—and not because the phrase is a failure but because it is a revelation—*Mules and Men* gestures toward the vision of posthumanist self-lessness that the conjure universe compels. It is a vision, finally, that unsettles but does not redeem or save "us." Yet, unlike *Eureka, The Education,* and *The Conjure Woman,* which also announce that we can neither refuse nor be the universe, *Mules and Men* and *Tell My Horse* do not claim that "we" are thereby condemned. They reveal instead that our alienating intimacy with otherness is also, impossibly, the very flesh of our becoming, a never-completed process that does not preclude a posthumanist "we" yet to be. And yet, as a lesson in impossibility, it may also be an impossible lesson. Like Poe, Adams, the slave autobiographers, Chesnutt, and, indeed, Hurston herself, we might not ever be able to fully know that "we" are not our own, never completely comprehend a universe without "us." If, though, we are somehow less mystified in the wake of voodoo's revelation, then it is because in it we might begin to concede the mystery rather than the mastery of our existence. When the "I" of "I don't know" is lost to itself, when an epistemological rupture widens to become an ontological abyss, then might we glimpse what the conjure cosmologies of Chesnutt and Hurston fleetingly reveal: the infinite contingency of our condition, the radical uncertainty of our selves. What Hurston says of dancing while possessed we might thus say of living—of becoming and unbecoming—in the voodoo universe: "I danced, I don't know how" (*Mules and Men*, 242).

Coda
"THE COSMOPOLITICAL PARTY"

ALTERNATE UNIVERSES, none our own. Poe's inrushing, panpsychic things collapsing, with us, into a fatal identity. Adams's ever-accelerating diffusion, the entropic thinning of self, world, and biography. Chesnutt's conjure metamorphoses, where becoming one with the environment necessitates becoming undone as oneself. Hurston's voodoo (dis)possession, given momentary form by an autodestructive ethnography whose echo sounds the possibility of a posthumanist collective. Though their geneses and genres are different, the realities they double not the same (transcendentalism and mesmerism; popular Darwinism and optimistic historiographies; post-Reconstruction racism and early conservationism; U.S. imperialism and primitivist Africanism, respectively), all present cosmoses in which "we"—separate, superior, masterful—do not exist. In these universes without "us," there is neither domineering detachment nor self-exalting unity, only the paradox of the human self coming out of, and then being decomposed by, impersonal and nonhuman cosmic processes.[1] Here constitutive-yet-alienating forces define our existence, converting "us" into emergent, ephemeral events rather than static, pregiven things.

The texts that I consider frequently depict this loss of self-ownership as literal death, but they make clear that it applies to life as well, as people are opened to "other" bodies, agencies, and becomings traversing the spectrum of sameness and difference, subject and object, inside and outside. Persons still inhabit these universes, then, but only insofar as they are, in turn, inhabited, the impermeable, abyssal dualisms that once guarded their ontological privilege now ceded to dynamic, co-prosthetic relations with things and energies no longer dependent on "us" for their vitality or meaning. Both more and less than previously imagined, the self—like a mesmerized patient, a dynamo, a riverbank, a voodoo altar—thus becomes an assemblage of "foreign" powers. In this way, Poe, Adams, Chesnutt, and Hurston picture the *Blockhead*'s fate as our own, making the conjunction with cosmic forces simultaneously the wonder and the horror of our being.[2] Difference without cleavage. Enmeshment without holism.[3] Identity as relation rather than as separation.

Rather than dismiss these cosmologies as empirically false or merely symptomatic of an underlying ideology or psychology, I have endeavored in the preceding chapters to explore them in the spirit of what might be called, following Alfred North Whitehead, Gilles Deleuze, and Isabelle Stengers, *speculative criticism*. In this approach, echoing the "what if?" of speculative fiction, these weird realities are taken seriously to juxtapose them against our own, bringing the world we inhabit into relief so as to denaturalize its—and our—givenness. Neither systematic nor axiomatic, these cosmologies nevertheless articulate consistent philosophical principles, quandaries, *and affects*, ones indissociable from their respective narrative forms, which all warp, buckle, and swoon under the strain of attempting to represent that which exceeds "us." Far from being a failure of communication, however, this stress may be seen as the very mark "of adventure, of speculation, of search[ing] for new ideas," that Whitehead defines as philosophy.[4]

In addition to situating each cosmology in its historical context, then, I also have suggested how they might be placed in conversation—sometimes complementary, sometimes antagonistic—with different strains of current posthumanist thought (Poe, univocity, and thing theory; Adams and inorganic evolutionism; Chesnutt and radical environmentalism; Hurston and phenomenologies of ecstatic experience). The power of these juxtapositions comes from the fact that my authors' cosmologies refuse the self-serving fantasies of both "Enlightenment" and "post-Enlightenment" conceptions of the world, both the idea that "we" lord over the universe and the notion that "we" may become it. The cosmologies' shared dominant affect of fear, through perhaps unjustifiable in a strict sense (Cary Wolfe notes that "posthumanist theory need not indulge . . . Foucauldian dystopianism"), testifies to this dual refusal; each text is suffused with an uncanny unease that forestalls our domestication of the world.[5] Viewed in this light, the cosmologies that I examine are themselves "nonhuman actors," voicing untimely meditations on a posthumanism understood not as a belated reaction or a new kind of being but as an inassimilable movement or remainder *immanent to* both anthropocentric exceptionalism and its sublime romantic correction.[6]

In Poe's and Adams's renditions, this evacuation of humanist subjectivity—the very subjectivity that Douglass and Equiano attempted to fill with themselves—is linked to the end of existence itself. Doggedly mechanistic and deterministic, their universes speed toward inevitable, apocalyptic conclusions (implosive and explosive, respectively), for which the death of Man is the most poignant index. Anthropocentrism's—arguably white privilege's—last, perverse gasp. This doesn't mean that their ghosts don't continue to haunt "us," only that they are dead to the possibility of things being otherwise. Theirs is a rebuke unto extinction, meaning that their cosmologies need only play out their fixed misanthropic principles to the bitter end.

Embodying the diasporic emphasis on the necessarily unfinished nature of becoming, Chesnutt's conjure and Hurston's voodoo offer more open-ended cosmoses. Although they, too, abolish the master self, it is effected through the *ongoingness* of the universe, not its end, through a plurality of eternally competing forces and not a rigid causal chain cut to a certain length. The hope-terror of a final completion is nowhere to be found. Undoubtedly, their cosmoses are in their own way deterministic, but here the future is underdetermined, connected to the present, if at all, by nonlinear ruptures and recurrences rather than a fixed and unswerving line. Another way to say this is that diasporic "determinisms" are contingencies that constantly must be created at all points—in Bruno Latour's formulation, "'universality' is . . . local"—as evidenced by the chanciness, the in-all-senses riskiness, of the events taking place within them: accidents and coincidences abound; reversals of fortune are the norm; and no act, no matter how powerfully forged, seems conclusive or final (in the case of Hurston's zombies, death itself is partially overcome).[7] Even the formal structures of Chesnutt's and Hurston's texts (serially episodic, largely noncumulative) manifest the provisionality of knowledge and existence in the conjure cosmos. Unlike *Eureka* or "The Tendency of History," therefore, which omnisciently forecast the end that is yet to come (thereby betraying the essential stasis of their ostensibly dynamic cosmoses), *The Conjure Woman* tales, *Mules and Men,* and *Tell My Horse* suggest the impossibility of colonizing the future because of the spontaneous *creativity* of cosmic processes.

If this potentiality is circumscribed in Chesnutt's tales by their antebellum settings and their author's rationalist skepticism, then Hurston's texts explore it more openly and consistently (before they, too, pull back in the end). Unlike in Chesnutt's pathos-infused tragedies, here voodoo's cancellation of our ontological privilege does not necessitate the obliteration of embodied persons; contact with voodoo can result in physical death, but only the demise of the metaphysical self is certain. Far from mutual exclusivity, therefore, the voodoo universe presumes the paradoxical simultaneity of people and impersonal forces, the perpetual dance of possession in which "our" bodies become the loci or confluence of myriad nonhuman agencies. Consequently, a new conception of "we" is allowed to emerge, one defined by performance and action rather than by fixed taxonomic being—modeling what might be called, adapting a phrase from Chris Castiglia, a "posthumanism without posthumans."[8] As we saw in the preceding chapter, the rub is that these actions—their causes and effects—are never entirely "our" own. Like "us," they are neither self nor other but plural compositions, their degrees of freedom bound by the collective constraints that make them possible. As a result, "we" have existence but not purity, power but not mastery; our persons and our potency are alike trans-actional and thus beyond "our" full command. Or, in the language of Latour's revision of actor–network theory, "I never act; I am always slightly surprised by what I do, by the chance to mutate, to change, and to bifurcate, the chance that I and the circumstances surrounding me offer to that which has been invited, recovered, welcomed. . . . Action is not about mastery."[9]

The question with which I close—the one toward which the book as a whole has been building—is whether this newly defined, posthumanist "we" offers anything other than proscription. In addition to exposing fraudulent, violent attempts to cast the cosmos either in our image or as our shadow, do my authors' cosmologies—and

Hurston's in particular—articulate an affirmative alternative? More specifically, is there a positive ethics or politics that follows from their interlineations of self and cosmos, something commensurate with Gayatri Chakravorty Spivak's "planetarity," Ursula Heise's "sense of planet," or William Connolly's "world of becoming"?[10] If so, is it, despite its almost unimaginable scope, actually realizable, livable? Does it present something other than what Hurston dismisses as mere "wishful thinking" about "millenn[ial] . . . justice in the absolute"?[11] Can ontological flatness be translated into a workable, nontotalizing, nonviolent unity of concern? In the absence of mastery, can nonhierarchical assemblages authorize real collective action? Can, in short, any "we" be equal to its utterance?[12]

As we shall see, it is possible that the answer to all these questions is no, which would not necessarily be a failure, properly understood (the denaturalization of ontological inequality need not result in the naturalization of positive alternatives). It could be that these cosmologies' power—indivisible from their limit—rests precisely in their incompatibility with our ethicopolitical attempts to domesticate them, whether because they are relentlessly ahuman (Poe, Adams, Chesnutt) or because they are necessarily underdetermined (Hurston). And yet we might still wonder: if "we" are defined by action—and if, as Hurston contends, "equality is as you do it and not as you talk it"—then what is to be done?[13] Who—or what—are "we" to be?

SUCH QUESTIONS RESONATE with ones posed in critical theory's recent "cosmological" turn, in which various attempts have been made to articulate "a democracy of objects" (Levi Bryant), "an *object-oriented* democracy" or "parliament of things" (Latour), "a political *ecology*" (Jane Bennett), or, simply, a "cosmopolitics" (Stengers).[14] Although differences among these and related positions abound, what is striking is the consistency of their political vocabulary; each not only imagines the implications of "including nonhumans in political theory" but also actively advocates for it.[15] But what could politics in this context mean? How to conceptualize the universe of things, much less legislate it? How can one speak of a parliament beyond human beings without evacuating the concept of its purchase? What is a democracy in the absence of a coherent, human *demos*?

Because his work informs my own theoretical sensibilities, and because it is the most familiar in the American academy, let us attend to these questions by briefly considering the example of Latour, particularly his more recent books *We Have Never Been Modern, The Politics of Nature,* and *Reassembling the Social.* Deemphasizing the Nietzschean focus on contests of strength found in such early texts as *The Pasteurization of France,* these later books pivot to a project of imagining the construction of a newly conceived polity, one that now includes nonhuman objects—historically denied full agency and being by the "modern Constitution"—in the "*progressive composition of one common world.*"[16] Latour explains that the establishment of this revolutionary "political ecology" requires constantly raising two separate yet mutually informing questions: "*how many are we*?" and "*can we live together?*"[17]

The first asks us to reflect on what we mean by "us" and "we," on the criteria for discriminating who counts and what doesn't, in order to move beyond our old humanist prejudices and embrace a more capacious definition that encompasses those things—"human" and "non-" (and the "non-" at the heart of the "human")—that affect others in some way.[18] Based on a "redistribut[ion] [of] speech" and "the capacity to act as a social actor" to the nonhuman world, it is an intentionally broad definition, one designed so as not to exclude from the outset.[19] If it makes a difference that makes a difference, then, for Latour, it is to be counted, meaning that ontological equality inheres in action itself rather than in a preset, circumscribed range of actors ("the social is a . . . type of circulation," and "by following circulations we can get more than by defining entities").[20] And if the constituents of "we" are defined by action, then so, too, is the "we" they comprise. Calling himself "a John Dewey disciple," Latour makes clear that his "we" is less a universal, Platonic ideal than a series of ad hoc assemblages *called into being* by the local exigencies of shared matters of concern.[21] In other words, "we" is an *emergent, provisional* network of ontologically flattened human and nonhuman agencies linked—across dizzying temporal phases and spatial scales—in a common activity, performance, alliance, controversy, or event (e.g., global warming, the *Columbia* explosion, the identification of the germ responsible for tuberculosis). Not unlike the identity-in-heterogeneity we have seen in Hurston's hoodoo, this thing that is "we" is a congressing, a connecting, a gathering, a verb rather than a noun. Consequently, rather than a once-and-for-all census, "how many are we?" must constantly be re-posed, as the "we" in question changes with the circumstance at hand (indeed, it *is* the circumstance). For any new event, a new "we." There is, Latour insists, always a "*multiplicity* of . . . new beings for which room must be made."[22]

Whereas the first question is methodological and empirical, its objective descriptive rather than prescriptive ("It's not a moral point"), the second—"*can we live together?*"—asks whether the newly recognized agencies thereby discovered can be reassembled into a rehabilitated (because no longer preformed or artificially exclusionary) society.[23] "Once you have extended the range of entities," Latour explains, "the new associations do not form a livable assemblage. This is where politics again enters the scene if we . . . define politics as the intuition that associations are not enough, that they should also be *composed* in order to *design one* common world."[24] In other words, once we have counted who "we" are, we must shape it in our preferred image and/or mold it to the task at hand. While acknowledging "the sheer difficulty of assembling collectives made of so many new members" (it's why he insists on the initial separation of his two questions), Latour believes we can do so, even making such compositional efforts not only the right and necessary culmination of sociology but also "the only scientific and political cause worth living for."[25] Furthermore, in an age of ecological catastrophe (largely brought about by our refusal to acknowledge the networks of which we have always been a part), "we scarcely have much choice" but to make the world itself a matter of concern and the cosmos of things our partners in the endeavor.[26] "We" must be universal. To fail to do so would mean that his reimagined House of Commons would fall, like the House of Usher, into the abyss between knowing the multiplicity of the cosmos and being able to act on it.

This wedding of an empirical imperative and a moral one raises two potential problems. The first is that Latour at least appears to suggest that the best way to reassemble the social is to quixotically extend the category of human subjectivity to "the parliament of things," as when he claims that "other circulating entities have to be granted back some rights of citizenry, so that they, too, can have a seat with the older members."[27] Elsewhere he announces in a similar vein a "Humanism Redistributed" that "invest[s] all quasi-objects with action, will, meaning, and even speech."[28] Given this rhetoric, it would seem that the goal is additive: to bestow the privileges, rights, and statuses that were once the exclusive province of humans ("the older members") more democratically but not to upset or worry them in any fundamental way. Such an intervention would represent only another step in the history of the West's progressive liberalization, embodied at one crucial moment by the slave autobiographies discussed in chapter 3.[29] His position would thus be continuous with, perhaps even inspired by, movements like the 1872 Cosmo-political Party (also known as the Equal Rights Party), which advocated for women's suffrage, among other things, and which was the first party to nominate a woman, Victoria Woodhull (free lover, spiritualist, suffragette), for president of the United States and an African American, Frederick Douglass (without his consent), as vice president (Figure 4).[30]

As Latour's critics have long complained, however, it's not clear how the liberal paradigm—which ideally addresses the unjust exclusion of conscious, intentional *humans* from equal participation in *human* institutions—can be applied in good conscience to the inclusion of *nonhuman* things. A few exceptional "higher" chordates, perhaps, but *anything* that makes a difference? Doesn't the principle of inclusion become so broad as to render the resulting collection meaningless? According to many commentators, the answer is yes: Latour's ideas must fail as political and sociological analyses because they are, in the words of an early, prescient review by Simon Schaffer, "marked by the heresy of hylozoism, an attribution of purpose, will and life to inanimate matter, and of human interests to the nonhuman."[31] Indeed, for many of these critics, Latour's abandonment of "the subject-object schema" is not only "unworkable" but an evasion of the "moral project" of humanistic inquiry.[32] It is merely the airy fantasy of a man whose social privilege ensures that he need not sully himself with Realpolitik. Though his work wears political trappings, such critics claim, the body politic within them ("not only made of people!") is either too divided or too phantasmatic to stand.[33]

Although there are aspects of Latour's presentation that invite the

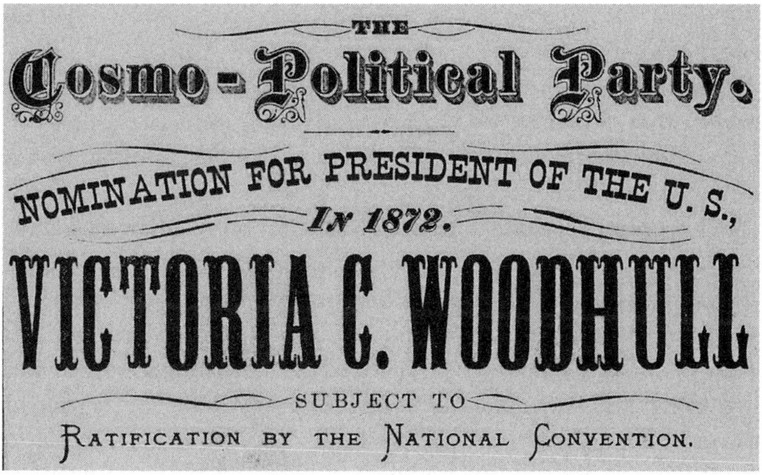

FIGURE 4. "The Cosmo-political Party, nomination for president of the U.S. in 1872. Victoria C. Woodhull." Schoenberg Center for Electronic Text and Image, University of Pennsylvania Libraries. Reproduced courtesy of the Library Company of Philadelphia.

criticism that he impossibly expands rather than interrogates liberal subjectivity, on the whole he undermines it for humans and nonhumans alike. This becomes clear when we realize that Latour is arguing not for the conferral of "humanity" to the nonhuman world but for the recognition that we have never been distinctly "human" ourselves—and thus are in no position to perpetuate it on our so-called others, which, for their part, have always been agents, have always possessed voices and dispositions blending (sometimes in amplification, sometimes in interference) with our own. For Latour (as for Michel Serres, Donna Haraway, Timothy Morton, and others), the received understandings of "human" and "nonhuman," or "self" and "other," are false purifications of profoundly complex entanglements. Agency cannot be assigned to one side and passivity to the other. Rather than transcendental dualisms, therefore, he proposes mutable, hybrid combinations, meaning that the "world" is no longer an external reality from which "we" are objectively detached but the field of relationality within which "we" are constituted; in this conception, the "human . . . is in the pass, in the sending, in the continuous exchange of forms.

Of course it is not a thing, but things are not things either."[34] This is not to say, then, that human difference is erased (Latour emphasizes that ANT does not represent "the dissolution of humanity"); it's only to note that the difference is a *perpetually* emergent one defined by its nonhuman relations.[35] If Latour's is a redistributed humanism, it is one stretched so far as to become unhuman.

The second objection to Latour's politicization of "interobjectivity" is less easily dismissed.[36] Specifically, it's not clear exactly how Latour's parliamentarianism adjudicates between a processual relationality that precedes entities, on one hand, and a procedural politics or ethics between entities, on the other. There seems to be an unfounded faith that things will work themselves out if we can only get everyone in the same room. To borrow the language of another debate, a politics of recognition does not ensure a politics of redistribution, and Latour fails to effectively bridge the two.[37] Once "the multiplicity of voices in [the] parliament of things is heard," what is to be done with them?[38] More to the point, who does the doing? Latour's reticence on this point is both principled and the consequence of a logical contradiction.

On one hand, he leaves the shape of the reassembled social undefined "to make sure there is no hegemony," to allow its constituents to form it in process rather than be beholden to an artificially preemptive standard; this is consistent with both his sociological method and his assertion that we humans are not masters of our actions or sole arbiters of what constitutes the public good ("we are not rabbis," "priests," "policemen," or "managers" charged with "put[ting] some order into the world").[39] On the other hand, however, Latour cannot specify what the nonmodern social should look like because to do so would expose the paradox at the heart of his politics—namely, that despite his emphasis on the ontological symmetry between humans and nonhumans in the constitution of the world (to the point where, as Stengers claims, it is "impossib[le] [to] giv[e] an adequate definition of" either in isolation), it is only the former who "look for ways to register the novelty of associations and explore how to assemble them in a satisfactory form."[40] It is a strictly human—perhaps even a professional—"us" that counts the "we" and decides on its politics. Yes, nonhumans have voices and agencies interlineated with our own, to which Latour rightly entreats us to be cognizant, but he implies that the ability to compose a common world that accommodates their heterogeneity depends on "us" (who else is asking his two questions, who else answering them?). Even if "we" must rely on "them" to effect the actual shaping, ours is the meta-agency.

As Latour well knows, once this distinction is granted, it is all too easy to act in the name of the others one supposedly merely follows. In our very desire to be responsible to the fact of interconnectedness, it is possible to reintroduce a form of discrete, exclusively human agency—even, indeed especially, when we put ourselves in a position of caring for the world or acting on its behalf. A benevolent dictatorship looms, the powers of which must be perpetually expanded to contain the resistances its rule creates.

Perhaps, then, we might hesitate to proceed from the first question to the second, following instead the "experimental metaphysics" of Poe, Adams, Chesnutt, and Hurston in *indefinitely* pausing before the awful fact of a "we" that can never be fully known or mastered because we (the we-that-is-I, the I's-that-are-we) can never be wholly identical to or apart from it.[41] Oscillating unto indistinction between objectivity and subjectivity, comprising movements both larger and smaller than selves, this "we"—like Vankirk of "Mesmeric Revelation," like Henry Adams, like Henry McAdoo of "The Goophered Grapevine," like Hurston—inhabits the in-between of presence and absence. Contingent, mutable, and yet thereby oddly durable, it is a "we" forged in the bonds of vulnerability and power, agency and passivity, that make and unmake "us," predate and survive "us."[42] It is an unceasing matter of concern, a living flesh, too strange, too close, too changeable to ever be converted into a dead matter of fact. Here, then, the never-to-be-answered question of who or what we are becomes so engrossing in its cosmological scope as to prevent the violence of our ever speaking or acting in its place.

To thus hesitate—physically, intellectually, affectively—before a reproblematized cosmos might seem a hopelessly naive or inadequate response to the immediate crises surrounding and comprising "us," but it is at least a failure that succeeds in forestalling the forms of mastery and certainty that precipitated them in the first place (like the governess in *The Turn of the Screw,* our assuredness has often proven lethal, from slavery to Manifest Destiny to social Darwinism to the operationalization of the world). In this sense, such a failure might even be considered a kind of care, one premised not on a patronizing control over—or separation from—our supposed inferiors (as in Heidegger's *Sorge*) but on a being-with expressed in remaining forever responsible to the bonds of our becoming. Illustrated in Julius's refusal to whip mules or kill frogs because they might be unknown relations and in Hurston's fleshly empathy with the brokenness of sacrificial animals, this withdrawal from harm is an embrace rather than a severance of relation. In fact, despite appearances to the contrary, it is not a covert expression of human power at all but an active constraint imposed by the ever more bodies—including our own—flooding the void between "us" and "not-us."[43] "We," too, are in need of care.

Thus, although Christopher Hitt is right to note that "embracing a sense of humility, mortality, and dependence" is not "a panacea" because "the consistent response of Western civilization . . . to [the] recognition of vulnerability has *not* been eventual acceptance, but dogged resistance," we can imagine a mode in which this constitutive uncertainty over our borders—this Adamic entropy, this Blockheadian erosion—is inhabited rather than overcome, thereby allowing our apparatuses of detached control to decay of their own accord.[44] Such an in-dwelling, what the previous chapter labeled *mutual possession,* does not require extinction, and it does not preclude action (indeed, it, like "we," *is* an action); it only invalidates mastery—only the notion that our actions (and our passivity) are unproblematically our own in either cause or effect. It involves only abstaining from exorcising the contingency, the ephemerality, the openness that first conjure and then haunt our becoming. Therefore, even if "the end of the ideal of omniscience" does not eventuate in "reenchantment," or love, or care, or any positive agenda whatsoever, it may yet ensure that "we," a local universe, cannot be naturalized or instrumentalized. Like Hurston's voodoo, it may yet dispossess us of ourselves.[45]

NOTES

NOTES TO INTRODUCTION

1 Begun circa 1519, the sculpture was part of a series of forty figures commissioned for the tomb of Julius II. The plan was never completed, however, and many figures, including the "slaves," "prisoners," or "captives" (of which the *Blockhead* is one), did not make it into the final design. See Ascanio Condivi, *The Life of Michelangelo*, 2nd ed., trans. Alice Sedgwick Wohl, ed. Helmutt Wohl (University Park: Penn State University Press, 1999); Charles De Tolnay, *The Tomb of Julius II* (Princeton, N.J.: Princeton University Press, 1954), vol. 4 of *Michelangelo*; and William Wallace, *Michelangelo: The Artist, the Man, and His Times* (Cambridge: Cambridge University Press, 1999). For the metaphysical context of the sculptor's work, see Michelangelo Buonarroti, *Complete Poems and Selected Letters of Michelangelo*, ed. Robert Linscott (Princeton, N.J.: Princeton University Press, 1980); Robert J. Clements, *Michelangelo's Theory of Art* (New York: New York University Press, 1961); and Tammy Smithers, "Michelangelo's Artistic Captivity as Mirrored in His Neoplatonic Captives," in *Neoplatonic Aesthetics: Music, Literature, and the Visual Arts*, ed. Liana De Girolami Cheney and John Hendrix, 211–26 (New York: Peter Lang, 2004).

2 Edith Balas, in *Michelangelo's Medici Chapel: A New Interpretation* (Philadelphia: American Philosophical Society, 1995), discusses the era's religious anthropocentrism, where man "is the representative of the universe and the essence of all its powers" (153).

3 Cameron's quotation comes from "'Avatar Friday': Fans Will Be Shown Preview of James Cameron's 3-D Film," *The Telegraph*, August 18, 2009. For an overview of the controversy that *Avatar* depicts a "white hero once again saving the primitive natives," see Jesse Washington, "'Avatar' Critics See Racist Theme," *Huffington Post*, January 11, 2010.

4 For studies that situate posthumanism transhistorically, see Caroline

Bynum Walker, *Metamorphosis and Identity* (New York: Zone Books, 2001); Jeffrey Jerome Cohen, *Medieval Identity Machines* (Minneapolis: University of Minnesota Press, 2003); Elaine L. Graham, *Representations of the Post/Human: Monsters, Aliens, and Others in Popular Culture* (New Brunswick, N.J.: Rutgers University Press, 2002); Keith Ansell Pearson, *Viroid Life: Perspectives on Nietzsche and the Transhuman Condition* (New York: Routledge, 1997); and Allison Muri, *The Enlightenment Cyborg: A History of Communications and Control in the Human Machine, 1660–1830* (Toronto: University of Toronto Press, 2007).

5 For definitions of transhumanism, see Nick Bostrom, "A History of Transhumanist Thought," *Journal of Evolution and Technology* 14, no. 1 (2005), http://jetpress.org/volume14/bostrom.html; Ray Kurzweil, *The Singularity Is Near: When Humans Transcend Biology* (New York: Viking, 2006); and the World Transhumanist Association homepage, http://www.transhumanism.org/. For humanist critiques of transhumanism, see Francis Fukuyama, *Our Posthuman Future: Consequences of the Biotechnology Revolution* (London: Profile Books, 2002), and Jürgen Habermas, *The Future of Human Nature* (Cambridge: Blackwell, 2003).

6 See Cameron's statement on the Home Tree Initiative website: "Today we find ourselves facing the greatest challenge of our time: saving our natural world from ourselves. . . . The time has come to stand up and be warriors for the Earth." http://www.avatarmovie.com/hometree/.

7 I draw here from Neil Badmington's "Theorizing Posthumanism," *Cultural Critique* 53 (Winter 2003): 10–27, which returns to Derrida's emphasis on the difficulty, if not the impossibility, of getting "outside" of humanism. For other critiques of the vestiges of humanism in much popular posthumanism, see N. Katherine Hayles's essential *How We Became Posthuman: Virtual Bodies in Cybernetics, Literature, and Informatics* (Chicago: University of Chicago Press, 1999); Cary Wolfe, *What Is Posthumanism?* (Minneapolis: University of Minnesota Press, 2010); Eugene Thacker, "Data Made Flesh: Biotechnology and the Discourse of the Posthuman," *Cultural Critique* 53 (Winter 2003): 72–97; and Daniel T. O'Hara's observations about posthumanism's very human "will-to-will" in "Neither Gods nor Monsters: An Untimely Critique of the 'Post/Human' Imagination," *boundary 2* 30, no. 3 (2003): 121.

8 Hayles, *How We Became Posthuman,* 287. In this light, the film disturbingly uses Jake's disability as a symbol of human dysfunction.

9 Bruno Latour, "An Attempt at a 'Compositionist Manifesto,'" http://www.bruno-latour.fr/articles/article/120-COMPO_MANIFESTO.pdf.

10 For two versions of Adorno's quotation, see Martin Jay, *Refractions of Violence* (New York: Routledge, 2003), 40, and Marshall Berman, *All That Is Solid Melts into Air: The Experience of Modernity* (New York: Verso, 1983), 126.

11 This differentiation between popular and critical posthumanisms borrows from ones previously made by Bart Simon, "Introduction: Toward a Critique of Posthuman Futures," *Cultural Critique* 53 (Winter 2003): 1–9, esp. 2–3, and Myra J. Seaman, "Becoming More (Than) Human: Affective Posthumanisms, Past and Future," *Journal of Narrative Theory* 37, no. 2 (2007): 246–75, esp. 247–50.

12 Donna J. Haraway, *Simians, Cyborgs, and Women: The Reinvention of Nature* (New York: Routledge, 1991), 178; Robert Pepperell, "The Posthuman Manifesto," *Kritikos* 2 (February 2005), http://garnet.acns.fsu.edu/~nr03/The%Posthuman%20Manifesto.htm; Daniel R. White, *Postmodern Ecology: Communication, Evolution, and Play* (Albany: State University of New York Press, 1998), 8; Paul Shepard, "Ecology and Man—A Viewpoint," in *The Subversive Science*, ed. Paul Shepard and D. McKinley (New York: Houghton Mifflin, 1969), 2, quoted in Neil Evernden, "Beyond Ecology: Self, Place, and the Pathetic Fallacy," in *The Ecocriticism Reader: Landmarks in Literary Ecology*, ed. Harold Fromm (Athens: University of Georgia Press, 1996), 102; Evernden, "Beyond Ecology," 101; Northrop Frye, *The Educated Imagination* (Toronto: CBC, 1961), 9, quoted in Evernden, "Beyond Ecology," 99. Evernden himself becomes skeptical of this position in his later *The Social Creation of Nature* (Baltimore: Johns Hopkins University Press, 1992).

13 Haraway, *Simians, Cyborgs, and Women*, 150, 152.

14 Andy Miah, "Posthumanism: A Critical History," http://ieet.org/archive/2007.04.12-MiahChapter2.pdf, 21. For the shift in ecological paradigms, see Evernden, "Beyond Ecology"; White, *Postmodern Ecology*; Daniel Botkin, *Discordant Harmonies: A New Ecology for the Twenty-First Century* (New York: Oxford University Press, 1992); Serpil Oppermann, "Theorizing Ecocriticism: Toward a Postmodern Ecocritical Practice," *ISLE* 13, no. 2 (2006): 103–28; and Dana Phillips, *The Truth of Ecology: Nature, Culture, and Literature in America* (New York: Oxford University Press, 2003).

15 Eleni Ikoniadou, "Rhythmic Topology: The Affective Stretching of Nature," in *An (Un)likely Alliance: Thinking Environment(s) with Deleuze/Guattari*, ed. Bernd Herzogenrath (Newcastle, U.K.: Cambridge Scholars, 2008), 173.

16 Tim Lenoir, "Makeover: Writing the Body into the Posthuman Tech-

noscape. Part One: Embracing the Posthuman," *Configurations* 10 (2002): 218.

17 Haraway, *Simians, Cyborgs, and Women,* 181, 154.

18 Miah, "Posthumanism," 21.

19 Bernd Herzogenrath, introduction to *An (Un)likely Alliance,* 4; Michael Mikulak, "The Rhizomatics of Domination: From Darwin to Biotechnology," in Herzogenrath, *An (Un)likely Alliance*, 68. Deleuze and Guattari themselves remark on this possibility in the introduction to *A Thousand Plateaus: Capitalism and Schizophrenia,* trans. and forward by Brian Massumi (Minneapolis: University of Minnesota Press, 1987), though the uncomfortable proximity between becoming imperceptible and becoming everything remains.

20 See Bruno Latour, *We Have Never Been Modern,* trans. Catherine Porter (Cambridge, Mass.: Harvard University Press, 1993).

21 See Isaac Newton, *Philosophical Writings,* ed. Andrew Janiak (Cambridge: Cambridge University Press, 2004); B. J. T. Dobbs, *The Foundations of Newton's Alchemy; or The Hunting of the Greene Lyon* (Cambridge: Cambridge University Press, 1975); and Frances Yates, *The Rosicrucian Enlightenment* (New York: Routledge Classics, 2002).

22 Francis Bacon, *New Atlantis,* ed. Alfred B. Gough (Oxford: Clarendon Press, 1915), 35.

23 See Marshall Brown, "Romanticism and Enlightenment," in *The Cambridge Companion to British Romanticism,* ed. Stuart Curran, 34–55 (Cambridge: Cambridge University Press, 2010).

24 See Iaian Hamilton Grant's discussion of Schelling's "metaphysics of nature" in *Philosophies of Nature after Schelling* (New York: Continuum, 2008), vii. My necessarily general account of romantic notions of self and world is based on M. H. Abrams's classic *Natural Supernaturalism: Tradition and Revolution in Romantic Literature* (New York: W. W. Norton, 1973); Richard Holmes, *The Age of Wonder: The Romantic Generation and the Discovery of the Beauty and Terror of Science* (New York: Vintage, 2010); *Romantic Science: The Literary Forms of Natural History,* ed. Noah Heringman (Albany: State University of New York Press, 2003); and Isaiah Berlin's lectures on (primarily) German romanticism collected in *The Roots of Romanticism* (Princeton, N.J.: Princeton University Press, 2001).

25 See Jonathan Bate's *Romantic Ecology: Wordsworth and the Environmental Tradition* (New York: Routledge, 1991), and Kate Rigby's *Topographies of the Sacred: The Poetics of Place in European Romanticism* (Charlottesville: University of Virginia Press, 2004).

26 Samuel Taylor Coleridge, "The Eolian Harp," in *The Great Romantics*

(New York: Quality Paperback Book Club, 1994), 579; William Wordsworth, "Influence of Natural Objects," in *The Great Romantics*, 110.

27 I draw here on Carolyn Merchant, *The Death of Nature* (London: Wildwood House, 1982); Val Plumwood, *Feminism and the Mastery of Nature* (London: Routledge, 1993); and, from a different perspective, Richard H. Grove, *Green Imperialism: Colonial Expansion, Tropical Island Edens, and the Origins of Environmentalism, 1600–1860* (New York: Cambridge University Press, 1995). In *The Song of the Earth* (Cambridge, Mass.: Harvard University Press, 2000), Bate revises his earlier stance by more critically examining Wordsworth's "Romantic egotism" (191), echoing Keats's diagnosis of Wordsworth's "egotistical sublime." Letter to Richard Woodhouse, October 27, 1818, in *Selected Letters of John Keats*, ed. Grant F. Scott (Cambridge, Mass.: Harvard University Press, 2002), 194.

28 Samuel Taylor Coleridge, "Religious Musings," in *Romanticism*, 2nd ed., ed. Duncan Wu (Oxford: Blackwell, 2000), 455.

29 Ralph Waldo Emerson, "The Over-Soul" (1847), in *Emerson: Essays and Poems* (New York: Literary Classics of the United States, 1996), 385–86; Walt Whitman, "Song of Myself" (1891–92), in *Walt Whitman: Leaves of Grass and Other Writings*, ed. Michael Moon (New York: W. W. Norton, 2002), lines 403–4, 406; Henry David Thoreau, "Walking," in *Thoreau: Collected Essays and Poems*, ed. Elizabeth Hall Witherell (New York: Literary Classics of the United States, 2001), 233.

30 Ralph Waldo Emerson, "Nature," in *Emerson: Essays and Poems*, 42.

31 Ralph Waldo Emerson, "Why Fear to Die," in *Emerson: Essays and Poems*, 1282.

32 Ralph Waldo Emerson, "The Method of Nature," in *Emerson: Essays and Poems*, 131. There is, of course, another Emerson who puts the individual self under the erasure of impersonal processes (this Emerson is not unrelated to the authors I study here). See chapters 3 and 4 of Sharon Cameron, *Impersonality: Seven Essays* (Chicago: University of Chicago Press, 2007); chapter 5 of Thomas M. Allen, *A Republic in Time: Temporality and Social Imagination in Nineteenth-Century America* (Chapel Hill: University of North Carolina Press, 2008); Branka Arsić, *On Leaving: A Reading in Emerson* (Cambridge, Mass.: Harvard University Press, 2010); and Arsić and Cary Wolfe, eds., *The Other Emerson* (Minneapolis: University of Minnesota Press, 2010).

33 Rudolf Steiner, *The Philosophy of Spiritual Activity*, in *The Philosophy of Spiritual Activity and Truth and Knowledge*, trans. Rita Stebbing, ed. Paul M. Allen (West Nyack, N.Y.: Rudolf Steiner, 1963), 122.

34 Marjorie Levinson, for instance, traces a common anthropocentric

attitude toward nature from the British romantics to posthumanist deep ecology: "For deep ecology, protection is nothing more than long-range and displaced production. Human stewardship objectifies nature in the sense of converting it conceptually if not literally into 'resources' that are valorized by long-term availability for development of a material or spiritual kind." Levinson, "Pre- and Post-dialectical Materialisms: Modeling Praxis without Subjects and Objects," in *Observing Complexity: Systems Theory and Postmodernity,* ed. William Rasch and Cary Wolfe (Minneapolis: University of Minnesota Press, 2000), 65.

35 For the constitutive tension in the United States between the desire to overcome nature and the desire to become it, where both result in forms of mastery, see Leo Marx's seminal *The Machine in the Garden: Technology and the Pastoral Idea in America* (Oxford: Oxford University Press, 2000), and Annette Kolodny, "Unearthing Herstory: An Introduction," in *The Ecocriticism Reader: Landmarks in Literary Ecology,* ed. Harold Fromm, 170–81 (Athens: University of Georgia Press, 1996).

36 Isabelle Stengers, *The Invention of Modern Science,* trans. Daniel W. Smith (Minneapolis: University of Minnesota Press, 2000), 34.

37 See Quentin Meillassoux, *After Finitude: An Essay on the Necessity of Contingency,* trans. Ray Brassier (New York: Continuum, 2008); Graham Harman, *Guerrilla Metaphysics: Phenomenology and the Carpentry of Things* (Peru, Ill.: Carus, 2005); and Ray Brassier, *Nihil Unbound: Enlightenment and Extinction* (New York: Palgrave Macmillan, 2007).

38 Jane Bennett, *Vibrant Matter: A Political Ecology of Things* (Durham, N.C.: Duke University Press, 2010). "By 'vitality,'" Bennett writes, "I mean the capacity of things—edibles, commodities, storms, metals—not only to impede or block the will and designs of humans but also to act as quasi-agents or forces with trajectories, propensities, or tendencies of their own" (viii). See also Bruno Latour, *Reassembling the Social: An Introduction to Actor-Network-Theory* (New York: Oxford University Press, 2005), 72: "There might exist many metaphysical shades between full causality and sheer inexistence."

39 See William Cronon, "The Trouble with Wilderness; or, Getting Back to the Wrong Nature," in *Uncommon Ground: Toward Reinventing Nature,* ed. William Cronon, 69–90 (New York: W. W. Norton, 1995); Molly Wallace, "'A Bizarre Ecology': The Nature of Denatured Nature," *ISLE* 7, no. 2 (2000): 137–53; and James C. McKusick, *Green Writing: Romanticism and Ecology* (New York: St. Martin's Press, 2000).

40 See Alfred North Whitehead, *Process and Reality,* ed. David Ray Griffin and Donald W. Sherburne (New York: Free Press, 1978), esp. 27–28,

55–56, 128–29, 214–15. In this I depart from Graham's object-oriented ontology, which makes relationality secondary to the withdrawn essences of objects. One problem with such a model is that it relies on a dyadic notion of relation rather than a pluralist one in which objects are composed—inside and out—of nearly infinite relations. For Graham's critique of the centrality that Whitehead and Latour grant to relation, see *Prince of Networks: Bruno Latour and Metaphysics* (Melbourne: re.press, 2009). For a rebuttal, see Steven Shaviro, "The Universe of Things," http://www.shaviro.com/Othertexts/Things.pdf.

41 See Michel Serres, *Hermes: Literature, Science, Philosophy,* ed. Josué V. Harari and David F. Bell (Baltimore: Johns Hopkins University Press, 1982), and *The Parasite,* trans. Lawrence R. Schehr (Minneapolis: University of Minnesota Press, 2007).

42 Pearson, *Viroid Life,* 1; Wolfe, *What Is Posthumanism?,* 186; Paul de Man, "Benjamin's 'The Task of the Translator,'" quoted in Barbara Johnson, *Persons and Things* (Cambridge, Mass.: Harvard University Press, 2010), 203. Against the emphasis in Lacan, Derrida, de Man, Johnson, Žižek, and Wolfe, I am interested in the material as well as the linguistic dimensions of the fact that "the 'human' turns out to be inhabited at its core by the *inhuman* Thing that resides at the heart of the human subject." Wolfe, *What Is Posthumanism?,* 198. For classic formulations of this idea, see Gregory Bateson, *Mind and Nature: A Necessary Unity* (Cresskill, N.J.: Hampton Press, 2002), and Francisco J. Varela, Evan Thompson, and Eleanor Rosch, *The Embodied Mind: Cognitive Science and Human Experience* (Cambridge, Mass.: MIT Press, 1993).

43 David Wills's excellent *Prosthesis* (Stanford, Calif.: Stanford University Press, 1995) discusses the inalienable nature of human prostheticity. For quasi-objects and -subjects, see Michel Serres and Bruno Latour, *Conversations on Science, Culture, and Time* (Ann Arbor: University of Michigan Press, 1995).

44 In addition to Latour's *We Have Never Been Modern* and Hayles's *How We Became Posthuman,* see also Cary Wolfe's claim that "we have never been human" in the context of his discussion of Michel Serres. Wolfe, "Bring the Noise: *The Parasite* and the Multiple Genealogies of Posthumanism," in Serres, *The Parasite,* xii: "For Serres, the posthuman precedes and subtends the human, both ontologically and epistemologically."

45 Note that Poe's title is "The Man That Was Used Up" rather than "The Man Who Was Used Up," a formal reflection of the story's thematic dehumanization.

46 Although the examples of such alternatives are varied, see in particular

Eve Kosofsky Sedgwick, *Touching Feeling: Affect, Pedagogy, Performativity* (Durham, N.C.: Duke University Press, 2003); Bruno Latour, "Why Has Critique Run Out of Steam? From Matters of Fact to Matters of Concern," *Critical Inquiry* 30, no. 2 (2004): 225–48; Michael Warner, "Uncritical Reading," in *Polemic: Critical or Uncritical,* ed. Jane Gallop, 13–38 (New York: Routledge, 2004); Rita Felski, *Uses of Literature* (Malden, Mass.: Blackwell, 2008); Stephen Best and Sharon Marcus, "Surface Reading: An Introduction," *Representations* 108 (Fall 2009): 1–21; Timothy Bewes, "Reading with the Grain: A New World in Literary Criticism," *differences* 21, no. 3 (2010): 1–33; and Toril Moi, "The Adventure of Reading: Literature and Philosophy, Cavell and Beauvoir," *Literature and Theology* 25, no. 2 (2011): 125–40.

47 Although Stengers locates the original break between human and nonhuman sciences with Galileo, she claims that the disciplining of "normal science" that excluded its resistors (including experimentally resistant nonhuman actors) was "created in the course of the nineteenth century with the institution of . . . modern academic research." Stengers, *Invention of Modern Science,* 117. Similarly, Latour places the origin of the modern divide in the seventeenth century but notes its intensification in the nineteenth and twentieth centuries. Latour, *Reassembling the Social,* esp. 83–85.

48 See Lorraine Daston and Peter Galison, *Objectivity* (New York: Zone Books, 2007).

49 Isabelle Stengers, *Cosmopolitics I,* trans. Robert Bononno (Minneapolis: University of Minnesota Press, 2010), 182, offers a methodology of "anamnesis" that restores the historical contingencies behind the supposedly inevitable triumph of objective rationality, which achieves its authority through the delegitimization of what it deems to be irrational: "the nineteenth century . . . put into place a . . . separation, to which present-day science owes its characteristics: the social separation between those who 'know how to recognize the facts' and those who are incompetent and only have opinions." "Complexity: A Fad?," in *Power and Invention: Situating Science,* trans. Paul Bains (Minneapolis: University of Minnesota Press, 1997), 18. See also Latour, *Reassembling the Social,* esp. 81–82, 89–90, 117–19, and Barbara Herrnstein Smith, *Scandalous Knowledge: Science, Truth, and the Human* (Durham, N.C.: Duke University Press, 2006).

50 For studies that demonstrate the extensive mutual engagement between U.S. literature and science at this time, see Mark Seltzer, *Bodies and Machines* (New York: Routledge, 1992); Laura Otis, *Networking:*

Communicating with Bodies and Machines in the Nineteenth Century (Ann Arbor: University of Michigan Press, 2001); Jane F. Thrailkill, *Affecting Fictions: Mind, Body, and Emotion in American Literary Realism* (Cambridge, Mass.: Harvard University Press, 2007); Priscilla Wald, *Contagious: Cultures, Carriers, and the Outbreak Narrative* (Durham, N.C.: Duke University Press, 2008); Paul Gilmore, *Aesthetic Materialism: Electricity and American Romanticism* (Stanford, Calif.: Stanford University Press, 2009); and, closest to my interests here, Eric G. Wilson, *Romantic Turbulence: Chaos, Ecology, and American Space* (New York: St. Martin's Press, 2000).

51 See Thomas Nagel, *The View from Nowhere* (New York: Oxford University Press, 1989).

52 Although calls to recognize the situatedness of observation are ubiquitous, Wolfe's *What Is Posthumanism?* and Bruce Clarke's *Posthuman Metamorphosis: Narrative and Systems* (New York: Fordham University Press, 2008) are the best discussions of its relevance to posthumanism.

53 Stengers, *Cosmopolitics I*, 37.

54 Edgar Allan Poe, "Ligeia," in *Poe: Poetry, Tales, and Selected Essays* (New York: Literary Classics of the United States, 2001), 265.

55 Edgar Allan Poe, *Eureka: A Prose Poem* (1848), in *Poe: Poetry, Tales, and Selected Essays*, 1345, 1353.

56 Ibid., 1358, 1349.

57 Ibid., 1314.

58 Ibid., 1353, 1357.

59 Edgar Allan Poe, "The Fall of the House of Usher," in *Poe: Poetry, Tales, and Selected Essays*, 328.

60 See Max Baym, *The French Education of Henry Adams* (New York: Columbia University Press, 1951), 20, 96.

61 Henry Adams, *The Education of Henry Adams*, ed. and with an introduction by Ernest Samuels (1918; repr., Boston: Houghton Mifflin, 1973), 6, 93, 109, 381.

62 Adams, *Education*, 457; Henry Adams, *A Letter to American Teachers of History* (1910), in *The Degradation of the Democratic Dogma* (New York: Macmillan, 1920), 166.

63 Stephen Crane, "The Open Boat," in *Stephen Crane: Prose and Poetry* (New York: Literary Classics of the United States, 1984), 907; Jack London, *The Sea-Wolf* (New York: Modern Library, 2000), 16. In *American Literature and the Universe of Force* (Durham, N.C.: Duke University Press, 1981), Ronald E. Martin includes Adams within the naturalist tradition.

64 Adams, *Education*, 382.

65 Russ Castronovo, *Necro Citizenship: Death, Eroticism, and the Public Sphere in the Nineteenth-Century United States* (Durham, N.C.: Duke University Press, 2001). Although Castronovo does not emphasize the connection, his book indirectly builds on Leslie A. Fiedler's classic *Love and Death in the American Novel* (Champaign, Ill.: Dalkey Archive Press, 1998).

66 Here and throughout, I use *voodoo* rather than the Haitian *Vodou* or the West African *Vodun* because it is the most frequent spelling in the United States during Chesnutt's and Hurston's careers and is also the spelling preferred by Hurston.

67 On the parallels between transcendentalism, spiritualism, New Thought, Theosophy, the modern New Age movement, and their popularized African American counterparts, see Jeffrey Anderson, *Conjure in African American Society* (Baton Rouge: Louisiana State University Press, 2005), and John Parker Deveney, *Paschal Beverly Randolph: A Nineteenth-Century Black American Spiritualist, Rosicrucian, and Sex Magician* (Albany: State University of New York Press, 1997).

68 See "Universal Beneficence," in John Burroughs's *Accepting the Universe: Essays in Naturalism* (Boston: Houghton Mifflin, 1920). This is a late essay, but the sentiment runs throughout Burroughs's career.

69 See Michael A. Elliott, *The Culture Concept: Writing and Difference in the Age of Realism* (Minneapolis: University of Minnesota Press, 2002).

70 Zora Neale Hurston, *Dust Tracks on a Road*, in *Zora Neale Hurston: Folklore, Memoirs, and Other Writings* (New York: Literary Classics of the United States, 1995), 712, 700.

71 See Pearson, *Viroid Life*, 155.

72 See Christopher Castiglia, *Interior States: Institutional Consciousness and the Inner Life of Democracy in the Antebellum United States* (Durham, N.C.: Duke University Press, 2008), and Gayatri Chakravorty Spivak, *Death of a Discipline* (New York: Columbia University Press, 2003), 71.

73 Johnson, *Persons and Things*, 240n1. See also ibid., 94, and Butler, *Giving an Account of Oneself* (New York: Fordham University Press, 2005).

74 See Theodor W. Adorno, *Negative Dialectics*, trans. E. B. Ashton (New York: Continuum, 2007), esp. 183–92. My thanks to Anne-Lise François for pointing out the relevance to my work of Adorno's attempts to circumvent the violence of forcing the natural world into anthropocentric schemas of identity. For recent posthumanist engagements

with Adorno, see Butler, *Giving an Account of Oneself*; Bennett, *Vibrant Matter*, 14–17; Wolfe, "In Search of Posthumanist Theory: The Second-Order Cybernetics of Maturana and Varela," in Rasch and Wolfe, *Observing Complexity*, 163–96; and Bill Brown, *A Sense of Things: The Object Matter of American Literature* (Chicago: University of Chicago Press, 2003), 18.

1. EDGAR ALLAN POE'S META/PHYSICS

1 Herbert Mayo, *Popular Superstitions, and the Truths Contained Therein, with an Account of Mesmerism* (Philadelphia: Lindsay and Blakiston, 1852), 10. In explaining away vampires as persons who actually are only in a "cataleptic state" owing to disruptions in their Od force, Mayo presents an account of a French woman presumed to be deceased but later revived by her lover. Interestingly, this story also appears in Poe's "The Premature Burial." Whether Mayo was directly familiar with Poe's tale, first published in 1844, or merely with Poe's own much older sources is indeterminable.

2 Prior to Mayo's nomination of the Od force, theories of the basis of mesmerism's power varied from invisible ethers to electromagnetic fluids, but all postulated that the underlying energy was universal. For the definitive history of mesmerism, see Alison Winter, *Mesmerized: Powers of Mind in Victorian Britain* (Chicago: University of Chicago Press, 1998).

3 See Mayo, *Popular Superstitions*, 169, 121. For the history of the displacement of mesmerism by hypnosis, see Léon Chertok and Isabelle Stengers, *A Critique of Psychoanalytic Reason: Hypnosis as a Scientific Problem from Lavoisier to Lacan* (Stanford, Calif.: Stanford University Press, 1992).

4 Mayo, *Popular Superstitions*, 23, notes that "terrestrial influence[s] [are] capable of disturbing the Od force in the body" as well as realigning it, and mesmerists often utilized items believed to have inherent mesmeric force (metallic objects, batteries, crystals, etc.). Franz Anton Mesmer, for instance, delivered his treatments using an iron scepter. For other examples of mesmerism by proxy, see Winter, *Mesmerized*.

5 For mesmerism in the United States, see Maria Tatar, *Spellbound: Studies on Mesmerism and Literature* (Princeton, N.J.: Princeton University Press, 1978); Robert Fuller, *Mesmerism and the American Cure of Souls* (Philadelphia: University of Pennsylvania Press, 1982); Bret Carroll, *Spiritualism in Antebellum America* (Bloomington: Indiana University

Press, 1997); and Samuel Coale, *Mesmerism and Hawthorne: Mediums of American Romance* (Tuscaloosa: University of Alabama Press, 1998).

6 Poe, for instance, knew the famous mesmerist and phrenologist Robert H. Collyer; met with (and later criticized) the Swedenborgian mesmerist Andrew Jackson Davis; sent letters to acquaintances indicating that "Mesmeric Revelation" was an exposition of his "own faith . . . somewhat detailed" (*The Letters of Edgar Allan Poe,* ed. John Ward Ostrom, 2 vols. [New York: Gordian Press, 1966], 1:259–60); and published an anonymous self-review of the same story that claimed it was "meant to be . . . the vehicle of the author's view concerning the DEITY" ("Edgar Allan Poe, *Tales,*" in *Edgar Allan Poe: Essays and Reviews* [New York: Literary Classics of the United States, 1984], 870). Thus, Poe's enjoyment over the fact that many people (including a few practicing mesmerists) were duped into thinking his tales were documentary case histories does not constitute a rejection of mesmerism itself. For fuller accounts of Poe's relation to mesmerism, see Sidney Lind, "Poe and Mesmerism," *PMLA* 62 (1947): 1077–94; Doris Falk, "Poe and the Power of Animal Magnetism," *PMLA* 84 (1969): 536–46; Hermann Schnackertz, "Mesmerizing the Reader," in *POEtic Effect and Cultural Discourses,* ed. Hermann Schnackertz, 117–62 (Heidelberg: Universitätsverlag Winter, 2003); Adam Frank, "Valdemar's Tongue, Poe's Telegraphy," *English Literary History* 72 (2005): 635–62; and Bruce Mills, *Poe, Fuller, and the Mesmeric Arts: Transition States in the American Renaissance* (Columbia: University of Missouri Press, 2006).

7 Walt Whitman, "One's-Self I Sing," lines 1, 6. In a similar vein, "Eidolons" pictures the poet ecstatically discovering his true identity—his "body permanent," "the real I myself"—within "the permanent life of life" that churns ceaselessly beneath the phenomenal world. Paul Gilmore's superb *Aesthetic Materialism: Electricity and American Romanticism* (Stanford, Calif.: Stanford University Press, 2009) surveys the ways electricity served as a dominant "figure . . . of subjective universality . . . [and] embodied transcendence" (5) for many canonical authors of American romanticism (including Whitman, Percy Shelley, and Emerson), not only in their literary works but also in their social and political imaginaries. My approach is indebted to his combination of aesthetic, philosophical, and political inquiry, and Poe might be understood as the macabre, pessimistic counterweight to the prevailing optimism and anthropocentrism of Gilmore's primary authors.

8 For the oft-noted convergence of liberal democracy and market capitalism in American individualism, see Steven Lukes, *Individualism* (Oxford: Basil-Blackwell, 1984).

9 For Poe's hostility toward nationalism in particular, see J. Gerald Kennedy, "'A Mania for Composition': Poe's Annus Mirabilis and the Violence of Nation-building," *American Literary History* 17 (2005): 1–35.

10 See Laura Dassow Walls's *The Passage to Cosmos: Alexander von Humboldt and the Shaping of America* (Chicago: University of Chicago Press, 2009) for a discussion of the influence of Humboldt's ecological thinking on Poe, particularly evident in *Eureka* (which is dedicated to Humboldt).

11 See Freud's discussion of the *unheimlich* as "a regression to a time when the ego had not yet marked itself off sharply from the external world" and as "something which is familiar and old-established in the mind [but] which has become alienated from it only through . . . repression." Freud, "The Uncanny," in *The Norton Anthology of Theory and Criticism,* ed. Vincent B. Leitch (New York: W. W. Norton, 2001), 941, 944. The *unheimlich* is thus related, conceptually and chronologically, to Freud's formulation of the *Todestrieb,* a drive to return to the inorganic. See Freud, *Beyond the Pleasure Principle* (New York: W. W. Norton, 1990).

12 See Charles O'Donnell, "From Earth to Ether: Poe's Flight into Space," *PMLA* 77, no. 1 (1962): 85–91; Evan Carton, *The Rhetoric of American Experience: Dialectic and Identity in Emerson, Dickinson, Poe, and Hawthorne* (Baltimore: Johns Hopkins University Press, 1985); and Joan Dayan, *Fables of Mind: An Inquiry into Poe's Fiction* (Oxford: Oxford University Press, 1987). See also Allan G. Halline, "Moral and Religious Concepts in Poe," *Bucknell University Studies* 2, no. 3 (1951): 126–50; Charles Feidelson Jr., *Symbolism and American Literature* (Chicago: University of Chicago Press, 1953); Richard P. Benton, ed., *Poe as Literary Cosmologer: Studies on Eureka: A Symposium* (Hartford, Conn.: Transcendental Books, 1975); Barbara Cantalupo, "*Eureka*: Poe's 'Novel Universe,'" in *A Companion to Poe Studies,* ed. Eric W. Carlson, 323–44 (Westport, Conn.: Greenwood Press, 1996); and Kenneth Alan Hovey, "Poe's Materialist Metaphysics of Man," in Carlson, *Companion to Poe Studies,* 347–66.

13 Edgar Allan Poe, "Letter to B—," in *Essays and Reviews,* 8. See also "The Rationale of Verse," in *Essays and Reviews*: "it is the nature of Truth in general, as of some ores in particular, to be richest when most superficial" (27).

14 Gilles Deleuze and Félix Guattari, *A Thousand Plateaus: Capitalism and Schizophrenia,* trans. Brian Massumi (Minneapolis: University of Minnesota Press, 1987), 151.

15 Edgar Allan Poe, "Mesmeric Revelation," in *Edgar Allan Poe: Tales and*

Sketches, ed. Thomas Ollive Mabbot, 2 vols. (Urbana: University of Illinois Press, 2000), 2:1033; hereinafter cited parenthetically by page number.

16 Poe favorably mentions Epicurus and Lucretius in multiple places in his nonfiction writing. He also briefly references Spinoza in a review of Rufus Dawes. Poe, "Rufus Dawes: A Retrospective Criticism," in *Essays and Reviews,* 495. For La Mettríe's argument that humanity is merely "a collection of springs" continuous with the rest of the universe, see his *Man a Machine* (La Salle, Ill.: Open Court, 1991), 135. La Mettríe's materialism dispenses with God, but Poe retains the idea as the singularity of undifferentiated matter; see his letter to Thomas Chivers on July 10, 1844 (the same year in which "Mesmeric Revelation" was published): "There is no such thing as spirituality. God is material. All things are material. Yet the matter of God has all of the qualities which we attribute to spirit: thus the difference is scarcely more than of words. There is a matter without particles—of no atomic constitution: this is God." Poe, *Letters,* 1:260.

17 Deleuze and Guattari, *A Thousand Plateaus,* 154, 507. Deleuze and Guattari take the BwO from a poem by Antonin Artaud, who named Poe as a primary influence; see *A New History of French Literature,* ed. Denis Hollier (Cambridge, Mass.: Harvard University Press, 1994), 723.

18 Michel Foucault, preface to Deleuze and Guattari, *Anti-Oedipus: Capitalism and Schizophrenia,* trans. Robert Hurley et al. (New York: Viking, 1977), xxi. *Anti-Oedipus* even includes a passing reference to Poe via Michel Carrouges (18).

19 Deleuze and Guattari, *A Thousand Plateaus,* 148.

20 Gilles Deleuze, *Francis Bacon: The Logic of Sensation,* trans. Daniel Smith (New York: Continuum, 2003), 27, 43, 27.

21 Ibid., 44; Gilles Deleuze, "Immanence: A Life," in *Pure Immanence: Essays on A Life,* trans. Anne Boyman (New York: Zone, 2001), 27.

22 Poe, *Letters,* 452, 362. Poe jealously guarded *Eureka* against misrepresentations by the press, even submitting a rebuttal of a critique that appeared in a minor New York literary weekly.

23 Dayan, *Fables of Mind,* 8, for instance, contends that to read Poe as actually arguing for cosmic unity rather than as savagely parodying it is to fall victim to an "all-too-human desire." See also Harold Beaver's classification of *Eureka* as a "hoax" in his introduction to *The Science Fiction of Edgar Allan Poe* (New York: Penguin, 1976), xvi, and Jonathan Elmer's more sophisticated situating of Poe within the "cultural

logic of the hoax" in *Reading at the Social Limit: Affect, Mass Culture, and Edgar Allan Poe* (Stanford, Calif.: Stanford University Press, 1995), 30.

24 Poe, *Eureka,* in *Edgar Allan Poe: Poetry, Tales, and Selected Essays* (New York: Library of America College Editions, 1996), 1259; hereinafter cited parenthetically by page number. Many critics claim that *Eureka* is either bad science that uses rhetorical flourishes to mask its logical deficiencies (Peter Swirski, *Between Literature and Science: Poe, Lem, and Explorations in Aesthetics, Cognitive Science, and Literary Knowledge* [Ithaca, N.Y.: McGill-Queen's University Press, 2000]) or an artwork more interested in pleasure than truth (Josef Jařab, "Edgar Allan Poe's Literary Strivings: How to Sell Beauty When Truth Is a Bore," in Schnackertz, *POEtic Effect and Cultural Discourses,* 51–60). *Eureka,* however, does not covertly present intuitions as facts (Michael Williams, *A World of Words: Language and Displacement in the Fiction of Edgar Allan Poe* [Durham, N.C.: Duke University Press, 1988]); it openly assumes that intuitions are equal, if not superior, to facts. More convincing, therefore, are the synthetic analyses of Paul Valéry's "On Poe's Eureka," in *Critical Essays on Edgar Allan Poe,* ed. Eric W. Carlson (Ann Arbor: University of Michigan Press, 1987), 102–7, and John T. Irwin's *American Hieroglyphics: The Symbol of the Egyptian Hieroglyphics in the American Renaissance* (New Haven, Conn.: Yale University Press, 1980).

25 Schelling advocated for a "necessary bond between philosophy and physics," as quoted in Iain Hamilton Grant, *Philosophies of Nature after Schelling* (New York: Continuum, 2008), ix.

26 Poe elaborates: "What you call The Universe . . . is but [God's] present expansive existence. . . . Those inconceivably numerous things which you designate as his creatures . . . are really but infinite individualizations of Himself" (*Eureka,* 1358). Unlike in "Mesmeric Revelation," however, here "an action of God [can] retur[n] upon itself" ("Mesmeric Revelation," 723).

27 *Eureka* characterizes gravity in the loaded language of "sympathy": "Each atom . . . sympathizes with the most delicate movements of every other atom" (*Eureka,* 1286). And the cause of this "sympathy" is a familial relation: "Does not so evident a brotherhood among the atoms point to a common parentage?" (*Eureka,* 1286–87). This is an anthropomorphism that troubles the terms of the pathetic fallacy by genuinely worrying humanity's claim to essential distinction. For the scientific sources on which Poe draws in *Eureka,* including Newton's gravity, see

Margaret Alterton's still useful *Origins of Poe's Critical Theory* (Iowa City: University of Iowa Press, 1925).

28 Even to study the universe, we must lose ourselves to it: "We need so rapid a revolution of all things about the central point of sight that, while the minutiae vanish altogether, even the more conspicuous objects become blended into one. . . . A man, in this view, becomes Mankind; Mankind a member of the cosmical family of Intelligences" (*Eureka,* 1262). Here the individual spins out of its identic limits, (r)evolving even out of "Mankind" to become, in effect, one of the "objects" under study, a blurred point within space rather than a detached recorder of it.

29 Edgar Allan Poe, "The Island of the Fay," in *Tales and Sketches,* 1:601; hereinafter cited parenthetically by page number. The "endowment of matter with vitality" ("Island of the Fay," 601) is central to the story: "We are madly erring, through self-esteem, in believing man, in either his temporal or future destinies, to be of more moment in the universe than that vast 'clod of the valley' which he tills and contemns, and to which he denies a soul for no more profound reason than that he does not behold its operation" ("Island of the Fay," 600–1). As in *Eureka,* which explicitly echoes this language, here is a panpsychic anthropomorphism that destabilizes, rather than confirms, "man."

30 Gilles Deleuze, "Whitman," in *Essays: Critical and Clinical,* trans. Daniel W. Smith and Michael A. Greco (New York: Verso, 1998), 58.

31 Irwin, *American Hieroglyphics*; Jonathan Auerbach, *The Romance of Failure: First-Person Fictions in Poe, Hawthorne, and James* (New York: Oxford University Press, 1989); Kenneth Silverman, *Edgar A. Poe: Mournful and Never-ending Remembrance* (New York: HarperCollins, 1992); Dennis Pahl, *Architects of the Abyss: The Indeterminate Fictions of Poe, Hawthorne, Melville* (Columbia: University of Missouri Press, 1989); and Carton, *Rhetoric of American Experience*, have all discussed how *Eureka* makes death a precondition for knowledge, but they do so to then offer either deconstructionist or psychoanalytic arguments that diverge from the point Poe is making here about the *material* basis of absolute identification.

32 Edgar Allan Poe, "The Conversation of Eiros and Charmion," in *Tales and Sketches,* 1:456, 459–60. Compare this with a passage from "Shadow—A Parable" (1835): "The peculiar spirit of the skies, if I mistake not greatly, made itself manifest, not only in the physical orb of the earth, but in the souls, imaginations, and meditations of mankind" (*Tales and Sketches,* 1:189).

33 Edgar Allan Poe, "The Power of Words," in *Tales and Sketches,* 2:1213.
34 Poe, "Conversation of Eiros and Charmion," 461.
35 Edgar Allan Poe, "The Colloquy of Monos and Una," in *Tales and Sketches,* 1:616–17.
36 Martin Heidegger, *Being and Time,* trans. Joan Stambaugh (Albany: State University of New York Press, 1996), 64, 68; hereinafter cited parenthetically by page number.
37 The conflation of the terms "humanity" and/or "self" with *Dasein* is not strictly fair to Heidegger's usage. However, insofar as Heidegger does present *Dasein* as the particular kind of being exclusive to humankind, the approximation will serve for the purposes of illustration.
38 For a precursor to Poe's thinking things, see Jonathan Lamb's discussion of eighteenth-century "it-narratives," which demonstrated how "the world of objects destabilize[d] personal identity." Lamb, "Modern Metamorphoses and Disgraceful Tales," *Critical Inquiry* 28 (Autumn 2001): 133.
39 See Graham Harman, *Tool-Being: Heidegger and the Metaphysics of Objects* (Peru, Ill.: Carus, 2002).
40 Against Heidegger's "conscious intention," Harman deploys his idea of the withdrawnness of things to counter Latour's claim that they are wholly deployed in relation. Graham Harman, *Prince of Networks: Bruno Latour and Metaphysics* (Melbourne: re.press, 2009), 140. Whitehead, conversely, imagines interconnection in language seemingly lifted from Poe: we are left "prey to vague terrors respecting a circumambient world. . . . In the dark there are vague presences, doubtfully feared. . . . The inflow into ourselves of feelings from enveloping nature overwhelms us; in the dim consciousness of half-sleep . . . we are left with the vague feeling of influences from vague things around us." Alfred North Whitehead, *Process and Reality* (New York: Free Press, 1978), 176.
41 Bill Brown, "Thing Theory," *Critical Inquiry* 28 (Autumn 2001): 5; Jane Bennett, *Vibrant Matter: A Political Ecology of Things* (Durham, N.C.: Duke University Press, 2010), xvi; Brown, "Thing Theory," 5. In *A Sense of Things: The Object Matter of American Literature* (Chicago: University of Chicago Press, 2003), Brown claims that late-nineteenth-century literature anticipated modernist object-based epistemologies in desiring to find "ideas in things," and it is possible to see Poe as an early, perverse exemplar of the phenomenon.
42 Brown, "Thing Theory," 7; Bennett, *Vibrant Matter,* xvii.
43 Later in his career, Heidegger put more emphasis on the etymology

of thing as a "gathering" and "we" as "the be-thinged," but he still believed that only certain things were capable of such "thinging." Martin Heidegger, "The Thing," in *The Object Reader*, ed. Fiona Candlin and Raiford Guins (New York: Routledge, 2009), 118, 122, 120.

44 Diana Coole and Samantha Frost, "Introducing the New Materialisms," in *New Materialisms: Ontology, Agency, and Politics* (Durham, N.C.: Duke University Press, 2010), 10. See also William E. Connolly, *A World of Becoming* (Durham, N.C.: Duke University Press, 2011).

45 Gilles Deleuze, *Difference and Repetition*, trans. Paul Patton (New York: Columbia University Press, 1994), 28. See also Deleuze, *The Fold: Leibniz and the Baroque*, trans. Tom Conley (New York: Continuum, 2006).

46 Echoing Stanley Cavell's "Being Odd, Getting Even: Threats to Individuality," in *Reconstructing Individualism: Autonomy, Individuality, and the Self in Western Thought*, ed. Thomas C. Heller et al., 278–312 (Stanford, Calif.: Stanford University Press, 1986), Irwin's *The Mystery to a Solution: Poe, Borges, and the Analytic Detective Story* (Baltimore: Johns Hopkins University Press, 1996) argues that Poe makes the self "an identity in difference . . . an always-about-to-be-accomplished evenness constituted by its being originally and essentially at odds with itself" (413). Unlike Cavell and Irwin, I contend that Poe makes this "identity in difference" not only a linguistic or psychological principle but also, and primarily, a cosmogony.

47 Typical for Poe, even the universe can't stay buried; *Eureka* suggests that the supposedly conclusive collapse of the cosmos back into oneness represents merely an episode in the eternal oscillation between God's diffusion and concentration: "the processes . . . will be renewed forever, and forever, and forever; a novel Universe swelling into existence, and then subsiding into nothingness, at every throb of the Heart Divine" (*Eureka*, 1356). Here the systole–diastole of the telltale universe guarantees reincarnation through the perpetual pulse of "Life—Life—Life within life." This does not mean, however, that individual identities survive. If anything, perpetual cosmic motion destabilizes individuality all the more because it disrupts the finality of the progression from self to God: with another "swelling into existence," "a new and perhaps totally different series of conditions may ensue" (*Eureka*, 1356).

48 W. C. Harris, "Edgar Allan Poe's *Eureka* and the Poetics of Constitution," *American Literary History* 12, nos. 1–2 (2000): 1–40. Harris builds on Donald Pease's claim that Poe is self-consciously an author of public identities in *Visionary Compacts: American Renaissance Writings in Cultural Context* (Madison: University of Wisconsin Press, 1987), 159.

49 For Heidegger's exposition of how hammers further the being of *Dasein*, see *Being and Time*, 64–65. See also Latour's critique of this moment: "Once the hammer has fallen, shattering the world into facts on one side and fetishes on the other, nothing can stop the dual question from being posed: did you construct the thing yourself or is it autonomous?" Bruno Latour, *Pandora's Hope: Essays on the Reality of Science Studies* (Cambridge, Mass.: Harvard University Press, 1999), 280.

50 Edgar Allan Poe, "The Black Cat," in *Tales and Sketches*, 2:849; hereinafter cited parenthetically by page number.

51 In a similar vein, Poe's "Instinct vs. Reason—A Black Cat" (1840) emphasizes the denigrating effect of animal intelligence on human exceptionalism: "While the self-love and arrogance of man will persist in denying the reflective power to beasts, because the granting it seems to derogate from his own vaunted supremacy, he yet perpetually finds himself involved in the paradox of decrying instinct as an inferior faculty, while he is forced to admit its infinite superiority. . . . Instinct . . . is perhaps the most exacted intellect of all. It . . . [is] the divine mind itself acting *immediately* upon its creatures." Poe, *Tales and Sketches*, 1:478.

52 Edgar Allan Poe, "The Imp of the Perverse," in *Tales and Sketches*, 2:1220, 1221.

53 Edgar Allan Poe, "Ligeia," in *Tales and Sketches*, 1:310; hereinafter cited parenthetically by page number. For the conventional interpretation, see Arthur Quinn, *Edgar Allan Poe: A Critical Biography* (Baltimore: Johns Hopkins University Press, 1998), 269–70, and J. Gerald Kennedy, "Poe, 'Ligeia,' and the Problem of Dying Women," in *New Essays on Poe's Major Tales*, ed. Kenneth Silverman, 113–29 (Cambridge: Cambridge University Press, 1998).

54 In stating that Ligeia's eyes are "more profound than the well of Democritus" ("Ligeia," 313), the narrator anticipates the epigraph to "A Descent into the Maelstrom": "The ways of God in nature, as in Providence, are not as *our* ways; nor are the models that we frame any way commensurate to the vastness, profundity, and unsearchableness of His works, *which have a depth in them greater than the well of Democritus*" (*Tales and Sketches*, 1:577). Like the narrator of "A Descent into the Maelstrom," Ligeia's husband loses himself by peering into an overfull abyss.

55 Relevant here is Melissa A. Orlie's discussion of Nietzsche's "will to power as an impersonal force within our lives rather than . . . a property of individuals." Orlie, "Impersonal Matter," in Coole and Frost, *New Materialisms*, 118.

56 Edgar Allan Poe, "The Fall of the House of Usher," in *Tales and Sketches*,

1:397–98, 401; hereinafter cited parenthetically by page number.

57 The idea that the physical impacts the "morale" is recurrent in Poe's writing, appearing in reference both to a bound collection of Longfellow's poetry—"the *physique* has had vast influence upon the *morale*" of the reader ("Henry Wadsworth Longfellow, *Poems on Slavery*" [1845], in *Essays and Reviews,* 761)—and to "scientific music," which Poe regards as "the triumph of the *physique* over the *morale* of music" ("Rationale of Verse," 34).

58 Compare the narrator's "monomania" in Edgar Allan Poe, "Berenice," *Tales and Sketches,* 1:211: "[It] consisted in a morbid irritability . . . [toward] the most ordinary objects of the universe." The narrator stresses that his "absorption" is a purely sensory fixation on an "*invariably frivolous*" object (212) such as Berenice's teeth, which appear to have "a sensitive and sentient power, and even when unassisted by the lips, a capability of moral expression" (216).

59 The first record of "panpsychism" in the *Oxford English Dictionary* is 1879, but the idea was circulating for some time previously, especially in romantic thought. See Mabbott's notes on the real published texts collected in the fictional Usher library ("Usher," 419–20, notes 13–25), as well as one book published soon after the tale: Gustav Fechner's *Nanna, or On the Mental Life of Plants* (1848). These texts did not present panpsychism as a threat to humanity but as a deepening of the beauty and profundity of nature. See also David Skrbina, *Panpsychism in the West* (Cambridge, Mass.: MIT Press, 2005).

60 See J. Gerald Kennedy's introduction to *A Historical Guide to Edgar Allan Poe* (New York: Oxford University Press, 2001), 3–17.

61 Compare a passage from another story about the fusion of persons and environments, "Eleonora": "the question is not yet settled . . . whether all that is profound—does not spring from some disease of thought—from *moods* of mind exalted at the expense of the general intellect" (*Tales and Sketches,* 1:638). In "The Pit and the Pendulum," "swoons" trigger "forbidden" "memories" that speak of "the gulf beyond" (*Tales and Sketches,* 1:682–83).

62 Where Scott Peeples makes the House an allegory for the collapse of *figuration* ("Poe uses the house to reflect upon literary structures"), I argue that the House's literal de-construction embodies, for Poe, an inevitable tendency of matter. Peeples, "Poe's 'Constructiveness' and 'The Fall of the House of Usher,'" in *The Cambridge Companion to Edgar Allan Poe,* ed. Kevin J. Hayes (Cambridge: Cambridge University Press, 2002), 180.

63 Edgar Allan Poe, "Nathaniel Hawthorne, *Twice-Told Tales*" (May 1842), in *Essays and Reviews,* 572.

64 Edgar Allan Poe, "The Poetic Principle" (1850), in *Essays and Reviews,* 77.

65 Edgar Allan Poe, "Marginalia—March 1846," in *Essays and Reviews,* 1383; Poe, "The Philosophy of Composition" (1846), in *Essays and Reviews,* 23.

66 For political readings of the deindividuating effect of sensation in Poe's work, see Elmer, *Reading at the Social Limit*; Louis A. Renza, *Edgar Allan Poe, Wallace Stevens, and the Poetics of American Privacy* (Baton Rouge: Louisiana State University Press, 2002); and Terence Whalen, *Edgar Allan Poe and the Masses: The Political Economy of Literature in Antebellum America* (Princeton, N.J.: Princeton University Press, 1999). All three authors discuss how Poe deployed sensationalistic writing to effect a homogeneous reading public. For interpretations closer to my own interests, see Schnackertz, "Mesmerizing the Reader," and David Leverenz, "Spanking the Master: Mind–Body Crossings in Poe's Sensationalism," in Kennedy, *A Historical Guide,* 95–128.

67 Edgar Allan Poe, "Marginal Notes: A Sequel to the 'Marginalia'" (1845), in *Essays and Reviews,* 1363.

68 Poe, "Philosophy of Composition," 14. Compare the following from "Maezel's Chess-Player," *Essays and Reviews*: "we can without difficulty conceive the *possibility* of so arranging a . . . mechanism, that upon starting it with the *data* of the question to be solved, it should continue its movements regularly, progressively, and undeviatingly toward the required solution" (1256).

69 Poe, "Edgar Allan Poe, *Tales,*" 872; Edgar Allan Poe, "Edward Lytton Bulwer, *Night and Morning*" (1841), in *Essays and Reviews,* 150. Compare the account given in Poe, "Philosophy of Composition": "Having chosen a novel, first, and secondly a vivid effect, I consider whether it can best be wrought by incident or tone" (13).

70 Poe, "Edward Lytton Bulwer," 152.

71 Edgar Allan Poe, "The American Drama" (1845), in *Essays and Reviews,* 367. The sentiment is echoed in one of Poe's private letters: "the Godhead is observ[able] [only] in its effects." Poe, *Letters,* 381.

72 Edgar Allan Poe, "Chapter of Suggestions" (1845), in *Essays and Reviews,* 1293.

73 Poe, "Edgar Allan Poe, *Tales,*" 872.

74 Edgar Allan Poe, "Byron and Miss Chaworth," in *Poetry, Tales, and Selected Essays,* 947.

75 Edgar Allan Poe, "Marginalia—April 1846," in *Essays and Reviews,* 1392.

76 Poe, "Philosophy of Composition," 19; Poe, "Poetic Principle," 93; my emphasis.

77 Poe, "Philosophy of Composition," 15. Elsewhere, Poe argues that "the *highest* order of the imaginative intellect is always preëminently mathematical; and the converse." Edgar Allan Poe, "Rufus W. Griswald" (1842), in *Essays and Reviews,* 549.

78 Edgar Allan Poe, "Marginalia—July 1846," in *Essays and Reviews,* 1398. For a version of the Foucauldian author-effect that makes Poe a "crystallized form" of social relations, see Eliza Richards, *Gender and the Poetics of Reception in Poe's Circle* (Cambridge: Cambridge University Press, 2004), 3.

79 Edgar Allan Poe, "Thomas Moore" (1840), in *Essays and Reviews,* 334; Poe, "Marginalia—May 1849," in *Essays and Reviews,* 1451. Benesch discusses Poe's "cybernetic" model of authorship that privileges machinic reassembly over the "produc[tion of] 'original' data." Klaus Benesch, *Romantic Cyborgs: Authorship and Technology in the American Renaissance* (Amherst: University of Massachusetts Press, 2002), 114. Extending Benesch's argument in a direction that departs from my own, James Berkley offers an affirmative account of "post-human mimesis," in which Poe overcomes the alterity of technological sublimity by incorporating it into his writing method. See Berkley, "Post-human Mimesis and the Debunked Machine: Reading Environmental Appropriation in Poe's 'Maelzel's Chess-Player' and 'The Man That Was Used Up,'" *Comparative Literature Studies* 41 (2004): 356–76.

80 Poe, "Marginalia—May 1849," 1451.

81 Paul Valéry, "On Poe from *Situation de Baudelaire,*" in *The Unknown Poe,* ed. Raymond Foye (San Francisco: City Lights Books, 1980), 115.

82 Throughout his critical writings, Poe offers crystals as examples of the beauty of natural "equality." Poe, "Rationale of Verse," 33.

83 Poe, "Poetic Principle," 75; Poe, *Letters,* 1:256.

84 See chapter 5 of Whalen, *Edgar Allan Poe and the Masses.*

85 Edgar Allan Poe, "Thomas Babington Macaulay," in *Essays and Reviews,* 323–24.

2. HENRY ADAMS'S HALF-LIFE

1 Henry Adams, *The Education of Henry Adams,* ed. Ernest Samuels (1918; repr., Boston: Houghton Mifflin, 1973), 6, 93, 313, 109, 381; hereinafter cited parenthetically by page number.

2 "A Law of Acceleration" is *The Education*'s penultimate chapter.

3 T. S. Eliot, "A Sceptical Patrician," *The Athenaeum,* May 23, 1919, 361–62. It should be noted that the subtitle "An Autobiography" was not present in the privately printed 1907 edition of *The Education,* only being added by the publishers in the 1918 posthumous edition.

4 Adams's determinism is in evidence as early as the 1860s, when he writes that "everything in this universe has its regular waves and tides.... The laws which govern animated beings will be ultimately found to be... the same with those which rule inanimate nature, and as I entertain a profound conviction of the littleness of our kind, and of the curious enormity of creation, I am quite ready to receive... a systematic conception of it all.... I look for regular tides in the affairs of man.... In every progression... nations move by the same process which has never been explained but is evident in the ocean and the air. On this theory I should expect at about this time, a turn which would carry us backward." Henry Adams, *The Letters of Henry Adams,* 6 vols., ed. J. C. Levenson, Ernest Samuels, Charles Vandersee, and Viola Hopkins Winner (Cambridge, Mass.: Harvard University Press, 1982–88), 1:395–96.

5 Henry Adams, *The Degradation of the Democratic Dogma* (New York: Macmillan, 1920), 129, 269, 186, 141, 154, 229. *The Tendency,* "The Rule," and *A Letter* are hereinafter cited parenthetically by page number. The "Editor's Preface" to *The Education* explicitly mentions *Letter* as an effort to represent "the last... chapters of the 'Education'... in another way which might be more intelligible to students" (xxviii). The connection is all the more significant because the preface, though dated September 1918 and signed Henry Cabot Lodge, actually was written by Adams in 1916, in preparation for *The Education*'s publication after his death. See Samuels's note in *Education,* 541n13.

6 See John Conder, *A Formula of His Own: Henry Adams's Literary Experiment* (Chicago: University of Chicago Press, 1970).

7 Or as *The Education* states a few pages later, "order was an accidental relation obnoxious to nature" (*Education,* 457). Although she ultimately frames it in psychological terms, N. Katherine Hayles offers an insightful analysis of Adams's "encounter with chaos." Hayles, "The Necessary Gap: Chaos as Self in *The Education of Henry Adams,*" in *Chaos Bound: Orderly Disorder in Contemporary Literature and Science* (Ithaca, N.Y.: Cornell University Press, 1990), 73.

8 Isabelle Stengers, *Cosmopolitics I,* trans. Robert Bononno (Minneapolis: University of Minnesota Press, 2010), 182.

9 Turgot was a French economist who postulated the existence of fixed

phases of historical progress. Comte's positivist model of evolutionary progress postulated advancing stages of human thought. Buckle explained historical progress through scientific laws and physical data such as environmental conditions.

10 William H. Jordy's *Henry Adams: Scientific Historian* (New Haven, Conn.: Yale University Press, 1963) remains the standard for contextualizing Adams's science. See also Ronald Martin, *American Literature and the Universe of Force* (Durham, N.C.: Duke University Press, 1981), and Clive Bush, *Halfway to Revolution: Investigation and Crisis in the Work of Henry Adams, William James, and Gertrude Stein* (New Haven, Conn.: Yale University Press, 1991). For materialist "science[s] of history" (1) in general and of nineteenth-century attempts at a "social physics" (73–74) in particular, see Marvin Harris's seminal *The Rise of Anthropological Theory: A History of Theories of Culture,* rev. ed. (Walnut Creek, Calif.: AltaMira Press, 2001).

11 Lewis Henry Morgan, *Ancient Society* (New York: Henry Holt, 1877), 19.

12 Adams specifies that "this popular understanding of Darwinism had little to do with Darwin" (*Letter,* 153), making "popular . . . Darwinism" merely symptomatic of "the superficial and self-complacent optimism which seems to veneer the surface of society" (*Letter,* 180).

13 Asa Gray's "Natural Selection Not Inconsistent with Natural Theology" (1860) even attempted to harmonize evolutionary and Genesis accounts of human development. Cultural evolutionists also frequently included references to providential design, such as Morgan's mention of progress being "the plan of the Supreme Intelligence." Morgan, *Ancient Society,* 554.

14 *A Letter* mentions "philosophies of history . . . inspired by" each of these figures (148).

15 Edward B. Tylor, *Primitive Culture,* 2 vols. (1871; repr., Cambridge: Cambridge University Press, 2010), 2:410. Against their more conservative counterparts, Tylor and Morgan at least believed that nonwhite races were, like white civilizations of the distant past, *capable* of progressing from "barbarism" to "civilization." Morgan, *Ancient Society,* 554.

16 See Giorgio Agamben, *The Open: Man and Animal* (Stanford, Calif.: Stanford University Press, 2004).

17 Tylor, *Primitive Culture,* 2:410. Harris's *Rise of Anthropological Theory* devotes a chapter to "The Rise of Racial Determinism" (80–107). See also Robert C. Bannister, *Social Darwinism: Science and Myth in Anglo-American Social Thought* (Philadelphia: Temple University Press,

1979), and Ivan Hannaford, *Race: The History of an Idea in the West* (Baltimore: Johns Hopkins University Press, 1996), 325–68.

18 Tylor, *Primitive Culture,* 2:410. See William Wasserstrom, *The Ironies of Progress: Henry Adams and the American Dream* (Carbondale: Southern Illinois University Press, 1984).

19 H. G. Wells, *The Time Machine* (New York: W. W. Norton, 2009), 65.

20 Jennifer L. Fleissner, "The Ordering Power of Disorder: Henry Adams and the Return of the Darwinian Era," *American Literature* 84, no. 1 (2012): 31–60. Adams isn't wedded to entropy as a final explanation (he concedes that future discoveries could alter our understanding), and he certainly doesn't find it to be "convenient" for teaching (*Letter,* 261), but he does offer it as the best working account for historical trends. When he warns of the consequences of fully integrating physics and history, therefore, it is not because he doubts their conjunction but because he fears the debilitating effects it will have on the university and society at large.

21 According to Adams, the only way for thought not "to fall . . . under the second law of thermodynamics" (*Letter,* 199) is to divide, with Descartes, "spirit" from "body" (*Letter,* 191); Adams refuses to do so. For a discussion of how the revelation of entropy reversed the optimism associated with the discovery of the first law of thermodynamics, see Tina Choi, "Forms of Closure: The First Law of Thermodynamics and Victorian Narrative," *English Literary History* 74, no. 2 (2007): 301–22.

22 See Stengers, *Cosmopolitics I,* esp. 205–18. As confirmation of his belief in entropic decline, Adams cites authorities who claim that evolutionists have failed to offer "proof of any rise in vital energy" (*Letter,* 175) and that, instead, all indications show the human to be a degenerate organism whose celebrated "Reason" is only a degraded form "of the energy earlier known as Instinct or Intuition" (*Letter,* 192). In this last claim, Adams explicitly borrows from Bergson while converting the idea of "universal becoming" (*Creative Evolution,* trans. Arthur Mitchell [New York: Random House, 1944], 396) into "mechanical dissolution" (*Letter,* 206).

23 For histories of the minority applying pessimistic evolutionary theory to social phenomena at the fin de siéclè, see Daniel Pick, *Faces of Degeneration: A European Disorder, c.1848–c.1918* (Cambridge: Cambridge University Press, 1993); Arthur Herman, *The Idea of Decline in Western History* (New York: Free Press, 1997); and Dana Seitler, *Atavistic Tendencies: The Culture of Science in American Modernity* (Minneapolis: University of Minnesota Press, 2008).

24 For Adams's deviation from reformist sociology, see Henry S. Kariel,

"The Limits of Social Science: Henry Adams's Quest for Order," *American Political Science Review* 50, no. 4 (1956): 1074–92.

25 A later letter argues that "psychology, physiology, and history will join in proving man to have as fixed and necessary development as that of a tree; and almost as unconscious." Quoted in J. C. Levenson, *The Mind and Art of Henry Adams* (Boston: Houghton Mifflin, 1957), 126.

26 Adams's *Mont Saint Michel and Chartres* (New York: Penguin, 1986) similarly contends that "mankind would not admit an anarchical . . . universe" (348).

27 Joseph Mindel, for instance, says of "The Rule" that its "errors of fact and terminology, the misconceptions, and the scientific naïveté are sometimes disconcerting." Mindel, "The Uses of Metaphor: Henry Adams and the Symbols of Science," *Journal of the History of Ideas* 26, no. 1 (1965): 91. Jordy claims that Adams's application of scientific principles to history "obviously misrepresents the scientific method." Jordy, *Henry Adams: Scientific Historian,* 15. See also Stephen Miller, "Henry Adams: The Confusions of Cosmic Pessimism," *Partisan Review* 67, no. 2 (2000): 261–69.

28 Levenson, *Mind and Art of Henry Adams,* 374. Howard Munford claims that Adams "cynically" "used the reasoning of science to reveal the inadequacy of science" so that he might "shock his readers into awareness of what could happen" if society continued to pursue a technocratic future that would never be "humanly satisfying." "The last thing [Adams] wanted his readers to do," Munford concludes, "was to take his theories solemnly or to assume that the author did." Mumford, "Henry Adams and the Tendency of History," *New England Quarterly* 32, no. 1 (1959): 89, 86, 80, 89, 81. The idea that Adams's scientific histories are parodic is pervasive. Keith Burich has devoted three articles to it, the most forceful being "'Our Power is Always Running Ahead of Our Mind': Henry Adams's Phases of History," *New England Quarterly* 62, no. 2 (1989): 163–86. See also Vern Wagner, *The Suspension of Henry Adams* (Detroit, Mich.: Wayne State University Press, 1969); John Carlos Rowe, "Henry Adams," in *The Columbia Literary History of the United States,* ed. Emory Elliott, 645–67 (New York: Columbia University Press, 1988), esp. 663; and T. J. Jackson Lears, *No Place of Grace: Antimodernism and the Transformation of American Culture 1880–1920* (New York: Pantheon Books, 1981).

29 R. P. Blackmur claims that Adams reengineers deterministic sciences that "mutilated the individual," "accept[ing] the materials . . . of twentieth-century thought [only] as the decorum for his poetry." Black-

mur, *Henry Adams,* ed. Veronica A. Makowsky (New York: Harcourt Brace Jovanovich, 1980), 25, 263. Levenson views Adams's theories as "fable[s] in formulas which . . . used metaphors from modern science." Levenson, *Mind and Art of Henry Adams,* 363. See also Robert F. Sayre, *The Examined Self: Benjamin Franklin, Henry Adams, Henry James* (Princeton, N.J.: Princeton University Press, 1964), esp. 130, 134; Melvin Lyon, *Symbol and Idea in Henry Adams* (Lincoln: University of Nebraska Press, 1970), esp. 3, 108; and Howard Horwitz, "*The Education* and the Salvation of History," in *New Essays on* The Education of Henry Adams, ed. John Carlos Rowe, 115–56 (New York: Cambridge University Press, 1996).

30 Lionel Trilling, "Adams at Ease," in *A Gathering of Fugitives* (Oxford: Oxford University Press, 1980), 126. Henry Steele Commager might have been the first to argue along these lines: "The indifference of a modern world to the existence of a Henry Adams was less humiliating when it was seen to be the inevitable consequence of scientific forces over whose operation man exercised no control." Commager, "Henry Adams" (1937), in *Critical Essays on Henry Adams,* ed. Earl N. Harbert (Boston: G. K. Hall, 1981), 57. See also Sacvan Bercovitch, *The American Jeremiad* (Madison: University of Wisconsin Press, 1978), 194; James Goodwin, "The Education of Henry Adams: A Non-person in History," *biography* 6, no. 2 (1983): 117–35; Brooks D. Simpson, *The Political Education of Henry Adams* (Lawrence: University of Kansas Press, 2001); and Herman, *Idea of Decline,* 162–64. For a reading that casts Adams's theories as reactions to his wife's suicide, see Margaret J. Brown, "Henry Adams: His Passage to Asia," in Harbert, *Critical Essays on Henry Adams,* 243–57.

31 William Dusinberre, *Henry Adams: The Myth of Failure* (Charlottesville: University of Virginia Press, 1980), 213, 221, 211.

32 Martin, *American Literature and the Universe of Force,* 138, 143, 144. James Young concurs, believing that Adams's scientific theories evidence a "seriously unsettled state of mind." Young, *Henry Adams: The Historian as Political Theorist* (Lawrence: University of Kansas Press, 2001), 209.

33 There are exceptions, of course. Conder, *Formula of His Own,* xii, asserts that Adams attempts to popularize the deterministic sciences of his day through various "literary experiments." See also William Decker, "A Martyr to the Disease of Omniscience," in *Henry Adams and the Need to Know,* ed. William Decker and Earl N. Harbert, 315–44 (Boston: Massachusetts Historical Society, 2005), and Caleb Smith,

"Bodies Electric: Gender, Technology, and the Limits of the Human, circa 1900," *Mosaic* 41, no. 2 (2008): 111–26. Smith's argument and mine explore similar territory, but where he believes that Adams ultimately "domesticate[s]" "the new age of electric technology . . . by the invocation of the . . . virgin" (124), I argue that Adams proves such humanization to be impossible.

34 Munford, "Henry Adams and the Tendency of History," 89; my emphasis; Blackmur, *Henry Adams*, 263; my emphasis. Rowe also believes that Adams endorses "transcendent individualism" and the "manifold, ironic, metamorphic self that is mythologized in our culture as Emersonian man." Rowe, "Henry Adams," 666–67.

35 Robert Shafer, "Henry Adams," *International Journal of Ethics* 30, no. 1 (1919): 54–56, 56–57. See also Yvor Winters's *In Defense of Reason* (Denver, Colo.: Allen Swallow, 1947), which argues that *The Education*'s determinism "can only lead to automatism in action, and automatism is madness" (407). Adams, of course, would agree.

36 I use *impersonal* here in the sense suggested by Sharon Cameron in *Impersonality: Seven Essays* (Chicago: University of Chicago Press, 2007)—as an often violent "unbinding from the personal manifested as an unbinding from the human" (xiii–xiv).

37 In showing "Adams" to be guided blindly by forces beyond his control, *The Education* seconds Adams's historiographic essays in rejecting the "Great Men" approach in favor of the "scientific" laws of historical causation promoted by contemporaries such as Taine and Buckle. See Horwitz, "*The Education* and the Salvation of History," 124–26, and Levenson, *Mind and Art of Henry Adams*, 117–89.

38 Henry Adams, *The Letters of Henry Adams (1892–1918)*, ed. Worthington Chauncey Ford (New York: Houghton Mifflin, 1938), 510.

39 "The omnipotence of circumstance" is a phrase from Adams's *The Life of Albert Gallatin*, quoted in Ernest Samuels, *Henry Adams: The Middle Years* (Cambridge, Mass.: Harvard University Press, 1965), 52.

40 Harold Kaplan extends this textual passivity into Adams's biographical profile: "Adams made an obvious point of avoiding action, whether in politics, business, or even the academy; thus he assumed his lifetime persona—that of a passive, spectatorial, ironic, and subtly, comprehensively, despairing man." Kaplan, *Power and Order: Henry Adams and the Naturalist Tradition in American Fiction* (Chicago: University of Chicago Press, 1981), 39.

41 For a study of spiritualist automatic writing published in the same year as *The Education*, see Arthur Conan Doyle's *The New Revelation* (1918; repr., Guilford, U.K.: White Crow Books, 2009). The first

example of surrealist automatic writing—André Breton and Philippe Soupault's *Les champs magéntiques* (1919)—appeared the following year. See Rachel Leah Thompson's linkage of the two traditions in "The Automatic Hand: Spiritualism, Psychoanalysis, Surrealism," *Invisible Culture: An Electronic Journal for Visual Culture* 7 (Spring 2004), http://www.rochester.edu/in_visible_culture.

42 Adams's *History of the United States* argues that historians eventually will study humanity "by the same methods with which [they] stud[y] the formation of a crystal." Quoted in Levenson, *Mind and Art of Henry Adams*, 186.

43 Jean-Luc Nancy, *Corpus*, trans. Richard A. Rand (New York: Fordham University Press, 2008), 19.

44 See Mark Seltzer, *Bodies and Machines* (New York: Routledge, 1992); Laura Otis, *Networking: Communicating with Bodies and Machines in the Nineteenth Century* (Ann Arbor: University of Michigan Press, 2001); William A. Cohen, *Embodied: Victorian Literature and the Senses* (Minneapolis: University of Minnesota Press, 2009); Paul Gilmore, *Aesthetic Materialism: Electricity and American Romanticism* (Stanford, Calif.: Stanford University Press, 2009); and Smith, "Bodies Electric," esp. 115–16.

45 *The Education* also labels Adams "a Branly coherer" (384) and designates his mind a "sort of magnet, mechanically dispersing its lines of force" (434), a "telephonic centre" (433), a device "for detecting radio waves" (651n13), and a "thinking-machine" (180). The most famous of such conflations is *The Education*'s illustration of Adams's "ego . . . effac[ing] itself . . . to become a manikin" (xxx). Bearing an uncanny resemblance to the human form, but stripped, like the *Blockhead*, of its ontological vestments, the mannequin is Adams's figure of the "human condition," "treated as though it had life" (xxx). See Joseph G. Kronick, "The Limits of Contradiction: Irony and History in Hegel and Henry Adams," *Clio: A Journal of Literature, History, and the Philosophy of History* 15, no. 4 (1986): 391–410, esp. 395.

46 Quentin Meillassoux, *After Finitude: An Essay on the Necessity of Contingency*, trans. Ray Brassier (New York: Continuum, 2008), 9. Insofar as I am extending it precisely to those areas it is meant to bracket—i.e., "the modes of relation between a living creature and its environment" (12)—my use of Meillassoux's terminology isn't consistent with his own. I think, however, that the tension is a productive one, especially to the extent that Adams puts pressure on the notion that "correlationism" (5) is a wholly human predicate.

47 It is appropriate, then, that what should be the most cathected moment

of *The Education,* Adams's discussion of his sister's death, almost immediately abstracts itself into a Wordsworthian-cum-Kelvinian analysis of the universe's fundamental "chaos": "Nature's gesture . . . took form then as a phantasm, a nightmare, an insanity of force. For the first time, the stage-scenery of the senses collapsed; the human mind felt itself stripped naked, vibrating in a void of shapeless energies, with resistless mass, colliding, crushing, wasting, and destroying what these same energies had . . . labored from eternity to perfect. Society became fantastic, a vision of pantomime with a mechanical motion. . . . Mont Blanc . . . looked to him what it was—a chaos of anarchic and purposeless forces" (*Education,* 288–89). The point isn't that Adams projects his tragedy onto the world; it's that the world punctures his illusions of order and, thus, tragedy.

48 Neal Aubrey regards embodiment as Adams's solution to the split between objective science and subjective experience. Aubrey, "Henry Adams and the History of Postmodernism," *Canadian Review of American Studies* 24, no. 2 (1994): 23–60. Carolyn Porter, however, sees it as a methodological failure in *Seeing and Being: The Plight of the Participant Observer in Emerson, James, Adams, and Faulkner* (Middletown, Conn.: Wesleyan University Press, 1981).

49 Adams, *Letters of Henry Adams (1892–1918),* 635; Adams, quoted in Ernest Samuels, *Henry Adams: The Major Phase* (Cambridge, Mass.: Harvard University Press, 1964), 74. These letters record Adams's grief following his wife's suicide, but, as with Adams's description of his sister's death, they are also occasions to comment on the universe's "chaos of anarchic and purposeless forces" (*Education,* 288).

50 Adams, *Letters of Henry Adams (1892–1918),* 476. Because it marks a self dictating its own absence, *The Education* is also like a suicide letter: Adams described autobiographies as ways of "tak[ing] your own life" or of not "be[ing] murdered by anyone except [oneself]." Adams, *Letters of Henry Adams (1892–1918),* 495; quoted in Dusinberre, *Henry Adams,* 215. But unlike suicide notes and legal wills, which are paradoxically futural and extend agency beyond death, *The Education* postulates will-lessness in life as in death.

51 Goodwin, "Education of Henry Adams," 120. Samuels, *Major Phase,* 58, calls *The Education*'s perspective "posthumous."

52 Manuel DeLanda, *Intensive Science and Virtual Philosophy* (New York: Continuum, 2002), 170, 153.

53 Manuel DeLanda, *A Thousand Years of Nonlinear History* (New York: Zone Books, 1997), 17.

54 Manuel DeLanda, *A New Philosophy of Science: Assemblage Theory and Social Complexity* (New York: Continuum, 2006), 48; DeLanda, *Thousand Years,* 15.

55 DeLanda, *Thousand Years,* 16. See also DeLanda, "Nonorganic Life," in *Incorporations (Zone 6),* ed. Jonathan Crary and Sanford Kwinter, 129–67 (New York: Zone Books, 1992). For a popularized presentation of Prigogine's theories, see Ilya Prigogine and Isabelle Stengers, *Order Out of Chaos: Man's New Dialogue with Nature* (New York: Bantam, 1984).

56 William James, "The Dilemma of Determinism," in *The Writings of William James: A Comprehensive Edition,* ed. John J. McDermott (Chicago: University of Chicago Press, 1977), 595.

57 DeLanda, *Thousand Years,* 21, 274, 274, 274, 271.

58 Elizabeth Grosz, *The Nick of Time: Politics, Evolution, and the Untimely* (Durham, N.C.: Duke University Press, 2004), 13, 8. See also Keith Ansell Pearson's specific dismissal of Adams's "dubious linear, rational, additive" model in favor of a truly dynamic theory of history in *Viroid Life: Perspectives on Nietzsche and the Transhuman Condition* (New York: Routledge, 1997), 170.

59 Grosz, *Nick of Time,* 2.

60 Henri Bergson, *Creative Evolution* (New York: Random House, 1944), 182.

61 James, "Dilemma of Determinism," 593.

62 Paul de Man, "The Rhetoric of Blindness: Jacques Derrida's Reading of Rousseau," in *Blindness and Insight: Essays in the Rhetoric of Contemporary Criticism,* 2nd ed. (Minneapolis: University of Minnesota Press, 1983), 105, 113.

63 Joseph Parkhurst, for instance, claims that Adams "exploit[s] the tropological nature of autobiography in order to fashion a mask" that "gives to posterity a public artifact . . . while withholding the author's self." Parkhurst, "The Manikin and the Memorial Bronze: The Figure of Defacement in *The Education of Henry Adams,*" *biography* 17, no. 2 (1994): 144, 145.

64 Horwitz, "*Education* and the Salvation of History," 138, 142.

65 Gregory S. Jay, *America the Scrivener* (Ithaca, N.Y.: Cornell University Press, 1990), 211, 234.

66 For the parallels between postmodernism and chaos theory, see N. Katherine Hayles, ed., *Chaos and Order: Complex Dynamics in Literature and Science* (Chicago: University of Chicago Press, 1991).

67 De Man, "Rhetoric of Temporality," 218, 219, 222. See also Kronick, "Limits of Contradiction," 407.

68 See Nancy Armstrong, *How Novels Think: The Limits of Individualism from 1719–1900* (New York: Columbia University Press, 2005).

69 De Man, "Autobiography as De-facement," in *The Rhetoric of Romanticism* (New York: Columbia University Press, 1984), 74.

70 Barbara Johnson, *Persons and Things* (Cambridge, Mass.: Harvard University Press, 2008), 13, 14. Extending de Man's claims, Johnson notes that autobiographies function like prosopopeia (182).

71 See Jacques Derrida, *Specters of Marx: The State of the Debt, the Work of Mourning, and the New International,* trans. Peggy Kaumf (New York: Routledge, 1994).

72 Søren Kierkegaard, *The Concept of Irony with Continual Reference to Socrates,* ed. and trans. Howard V. Hong and Edna H. Hong (Princeton, N.J.: Princeton University Press, 1988), 152, 6; hereinafter cited parenthetically by page number.

73 Ayon Roy traces this self-collapsing irony to Schlegel, notes Hegel's critique of it as "mask[ing] an absolute subjectivity," and then charts Kierkegaard's "amplif[ication]" of Hegel's alternative: "'infinite absolute negativity.'" Roy, "Hegel contra Schlegel; Kierkegaard contra de Man," *PMLA* 124, no. 1 (2009): 117, 119. See also Mark C. Taylor, *Journeys to Selfhood: Hegel and Kierkegaard,* 2nd ed. (New York: Fordham University Press, 2000), and Kronick, "Limits of Contradiction."

74 De Man echoes Kierkegaard's assertion in *Concept of Irony,* 26, that "irony [is] infinitely silent" when he claims that the irony of our situation "eternally deprive[s us] of voice and condemn[s us] to muteness." De Man, "Autobiography as De-facement," 80.

75 Adams, *Letters of Henry Adams (1892–1918),* 445.

76 See Samuels, *Major Phase,* 62–63.

77 Adams, *Letters,* 4:430; Gustave Kobbe, "Mystery of Saint Gaudens' Masterpiece Revealed by John La Farge," *New York Herald,* January 16, 1910; Adams, *Letters,* 6:483. See also Adams, *Letters,* 5:488–89, 6:108, 109, 198 (and the note containing Theodore Roosevelt's description of the figure as "non-human" in *Letters,* 6:198–99n2); Cynthia R. Field, "The Adams Memorial," *Smithsonian Preservation Quarterly* (Summer/Fall 1995): 2–3; Homer St. Gaudens, *The Reminiscences of Augustus Saint-Gaudens* (New York: Century, 1913), esp. 356–66; and Parkhurst, "Manikin and the Memorial Bronze."

78 See Adams, *Letters,* 4:461, 462, 727–28, and Field, "Adams Memorial," 3. Although neither of the two Slaves whom Adams viewed in the Louvre was the *Blockhead,* they were part of the same series. Of another of Michelangelo's works, Adams writes, in language perfectly suited to the *Blockhead,* that it was "a mere formless effort, impossible

to complete, useless to define, furious to think about; a mere blind, brute endurance, without aim or hope or understanding; a primal energy, not even bound or aspiring." Adams, *Letters,* 4:727. For this reason, Adams names Michelangelo "chief of my school of Conservative Christian Anarchists"—a label Adams applies to himself in *The Education*—even though Michelangelo was not "much of a Christian, or a Conservative." Adams, *Letters,* 4:727.

79 *The Education* favorably cites Lucretius's *De Rerum Natura* twice (*Education,* 384, 459). For a comparison of Adams's Virgin and Lucretius's Venus, see Paul Hamill Jr., "The Future as Virgin: A Latter Day Look at the Dynamo and the Virgin of Henry Adams," *Modern Language Studies* 3, no. 1 (1973): 8–12.

80 See Wai Chee Dimock, *Through Other Continents: American Literature across Deep Time* (Princeton, N.J.: Princeton University Press, 2006); Ray Brassier, *Nihil Unbound: Enlightenment and Extinction* (New York: Palgrave Macmillan, 2007); Mark McGurl, "The Posthuman Comedy," *Critical Inquiry* 38, no. 3 (2012): 533–53; and Claire Colebrook, "Extinct Theory," in *Theory after "Theory,"* ed. Jane Elliott and Derek Attridge, 62–71 (New York: Routledge, 2011).

81 Colebrook, "Extinct Theory," 65.

82 Ibid., 69.

83 This perhaps explains the long-standing debate over Adams's position on U.S. imperialism. Goodwin and Kaplan agree that racism is the outcome of Adams "making humanity abstract" (Kaplan, *Power and Order,* 76), and Rowe contends that *The Education* can be interpreted as a "mystif[ying]" "code" that tacitly supports John Hay's neoimperialistic policies in his introduction to *New Essays on* The Education, 17, 18. Conversely, Daniel L. Manheim, echoing Herman, notes the "number of conflicted views" in Adams's thinking on imperialism. Manheim, "The Voice of Arii Taimai: Henry Adams and the Challenge of Empire," *Biography* 22, no. 2 (1999): 213–14.

84 Of Paul Bové's many excellent publications on Adams, see especially "Abandoning Knowledge: Disciplines, Discourse, Dogma," *New Literary History* 25 (1994): 601–19, and "Giving Thought to America: Intellect and *The Education of Henry Adams,*" *Critical Inquiry* 23, no. 3 (1996): 80–108.

85 See Thomas Carl Wall's *Radical Passivity: Levinas, Blanchot, and Agamben* (Albany: State University of New York Press, 1999): "radical passivity" "is passive with regard to *itself,* and thus . . . submits to itself as though it were an external power" (1).

3. "BY AN ACT OF SELF-CREATION"

1 See Giorgio Agamben, *Homo Sacer: Sovereign Power and Bare Life,* trans. Daniel Heller-Roazen (Stanford, Calif.: Stanford University Press, 1998).

2 Building on W. E. B. Du Bois's *The Negro Church* (1903; repr., Walnut Creek, Calif.: Alta Mira Press, 2003), Melville Herskovits's *The Myth of the Negro Past* (1941; repr., Boston: Beacon Press, 1958), and Albert J. Raboteau's *Slave Religion: The Invisible Institution in the Antebellum South* (New York: Oxford University Press, 1978), the scholarly consensus is now in favor of significant retentions of *various* African religious beliefs among U.S. slaves. Regarding the transatlantic continuities of conjure and voodoo–Vodun in particular, see Robert Farris Thompson, *Flash of the Spirit: African and Afro-American Art and Philosophy* (New York: Random House, 1983); Joseph M. Murphy, *Working the Spirit: Ceremonies of the African Diaspora* (Boston: Beacon Press, 1994); Anthony B. Pinn, *Varieties of African-American Religious Experience* (Minneapolis, Minn.: Fortress Press, 1998); Will Coleman, *Tribal Talk: Black Theology, Hermeneutics, and African/American Ways of "Telling the Story"* (University Park: Pennsylvania State University Press, 2000); Yvonne P. Chireau, *Black Magic: Religion and the African-American Conjuring Tradition* (Berkeley: University of California Press, 2003); and *Africanisms in American Culture,* ed. Joseph Holloway (Bloomington: Indiana University Press, 2005).

3 John Mbiti cautions that "there are about three thousand African peoples (tribes), and each has its own religious system." Mbiti, *African Religions and Philosophy* (Portsmouth, N.H.: Heinemann, 1990), 1. On this diversity, see also V. Y. Mudimbe, *The Invention of Africa: Gnosis, Philosophy, and the Order of Knowledge* (Bloomington: Indiana University Press, 1988), and Sidney Mintz and Richard Price, *The Birth of African-American Culture: An Anthropological Perspective* (Boston: Beacon, 1992).

4 Curtis J. Evans, *The Burden of Black Religion* (New York: Oxford University Press, 2008), 10.

5 In the cosmologies of both West Africa and the black American South, Will Coleman contends, the "demarcation between animate and inanimate, human and nonhuman, spirit and human, is almost non-existent; it is extremely fluid." Coleman, *Tribal Talk,* 25. See also Thompson, *Flash of the Spirit,* 117–45; Henry Mitchell, *Black Belief: Folk Beliefs in America and West Africa* (New York: Harper and Row, 1975), 72; and Lawrence Levine, *Black Culture and Black Consciousness:*

Afro-American Folk Thought from Slavery to Freedom (New York: Oxford University Press, 1977), 58.

6 See Mbiti, *African Religions and Philosophy,* 24, and Dominique Zahan, *The Religion, Spirituality, and Thought of Traditional Africa,* trans. Kate Martin and Lawrence Martin (Chicago: University of Chicago Press, 1979), 9.

7 Benjamin C. Ray notes that "in contrast to the Western notion of the soul as the 'essence' of man, the Dogon [and Yoruba] think of man in terms of multiple 'souls' or 'selves.'" Ray, *African Religions: Symbol, Ritual, Community* (Englewood Cliffs, N.J.: Prentice Hall, 1976), 134. Zahan concurs: "the individual does not constitute a closed system in opposition to the outside world." Zahan, *Religion, Spirituality, and Thought of Traditional Africa,* 9. For U.S. analogues, see Jeffrey E. Anderson, *Conjure in African American Society* (Baton Rouge: Louisiana State University Press, 2005), 37, and James Haskins, *Voodoo and Hoodoo: Their Tradition and Craft as Revealed by Actual Practitioners* (New York: Stein and Day, 1978), 27–28.

8 "During the height of spirit possession," Mbiti asserts, "the individual in effect loses his own personality and acts in the context of the 'personality' of the spirit possessing him." Mbiti, *African Religions and Philosophy,* 80–81. See also Geoffrey Parrinder, *West African Psychology: A Comparative Study of Psychological and Religious Thought* (London: Lutterworth Press, 1951), 181; Alfred Metraux, *Haiti: Black Peasants and Voodoo* (New York: Universe Books, 1960), 86; and Sheila Walker, *Ceremonial Spirit Possession in Africa and Afro-America: Forms, Meanings, and Functional Significance for Individuals and Social Groups* (Leiden, Netherlands: E. J. Brill, 1972).

9 Chireau, *Black Magic,* 106. For reported cases of this phenomenon, see Newbell Niles Puckett's *Folk Beliefs of the Southern Negro* (1926; repr., New York: Negro Universities Press, 1968), 249–55, and "Root Doctor: A Folklore Study in Which a Negro Healer Explains His Work before a Concealed Microphone to Harry M. Hyatt," Harry M. Hyatt Collection, Library of Congress, Washington, D.C.

10 See, among numerous others, Chireau, *Black Magic,* and Anderson, *Conjure.*

11 Christopher C. Fennell, *Crossroads and Cosmologies: Diasporas and Ethnogenesis in the New World* (Gainesville: University Press of Florida, 2007), 9.

12 Bruno Latour, *Reassembling the Social: An Introduction to Actor-Network-Theory* (New York: Oxford University Press, 2005), 45.

13 This is consistent with Latour's imperative to follow actors themselves,

thereby replacing hermeneutics with an "empirical metaphysics." Ibid., 51.

14 Ray, *African Religions,* 134.

15 Paul Gilroy, *The Black Atlantic: Modernity and Double Consciousness* (New York: Verso, 1993), 7–8, 59.

16 Ibid., 55, 58.

17 Frederick Douglass, *My Bondage and My Freedom,* in *Frederick Douglass: Autobiographies,* ed. Henry Louis Gates Jr. (New York: Library of America, 1994), 286; Gilroy, *Black Atlantic,* 60.

18 Gilroy, *Black Atlantic,* 68, 62, 68.

19 Douglass, *My Bondage,* 280. Douglass notes that Sandy "professed to believe in a system for which I have no name" (280). As Gilroy assumes, this "system" is most likely "conjure."

20 Gilroy, *Black Atlantic,* 60, 62, 60–61, 56.

21 Douglass, *My Bondage,* 281, 283. For other readings of the roots' failure, see Houston Baker Jr., *Blues, Ideology, and Afro-American Literature: A Vernacular Theory* (Chicago: University of Chicago Press, 1984), 46–47, and Barbara Johnson, *Persons and Things* (Cambridge, Mass.: Harvard University Press, 2008), 136–37.

22 See Robert S. Levine, *Martin Delany, Frederick Douglass, and the Politics of Representative Identity* (Chapel Hill: University of North Carolina Press, 1997), 99–143; Krista Walker, "Trappings of Nationalism in Frederick Douglass's 'The Heroic Slave,'" *African American Review* 34, no. 2 (2000): 233–47; Joanna Brooks, *American Lazarus: Religion and the Rise of African-American and Native American Literatures* (Oxford: Oxford University Press, 2003), 3, 8; and Michelle Wright, *Becoming Black: Creating Identity in the African Diaspora* (Durham, N.C.: Duke University Press, 2004), 10–11. Gilroy concedes that "the revolts of enslaved people made selective use of the ideologies of the western Age of Revolution." Gilroy, *Black Atlantic,* 44.

23 Douglass, "Fourth of July Oration," in *African Intellectual Heritage: A Book of Sources,* ed. Molefi Kete Asante and Abu S. Abarry (Philadelphia: Temple University Press, 1996), 640.

24 Joan Dayan, *Haiti, History, and the Gods* (Berkeley: University of California Press, 1995), 205.

25 Some even have claimed that this insistence on the race's humanity characterizes all of African American literature. See Carlyle Fielding Stewart III, *Soul Survivors: An African-American Spirituality* (Louisville, Ky.: Westminster John Knox Press, 1997); Jane Campbell, *Mythic Black Fiction: The Transformation of History* (Knoxville: University of

Tennessee Press, 1986); and James H. Cone, *Black Theology and Black Power* (New York: Seabury Press, 1969), 10.

26 Henry Louis Gates Jr., *The Signifying Monkey: A Theory of African American Literary Criticism* (Oxford: Oxford University Press, 1989), 128, 131, 165; Charles T. Davis and Henry Louis Gates Jr., eds., introduction to *The Slave Narrative* (New York: Oxford University Press, 1985), xxiii. See also Leonard Cassuto: "Slave narratives recount a key moment of transformation from human/thing to person, a transition accomplished during the movement—in thought and act—from slavery to freedom." Cassuto, *The Inhuman Race: The Racial Grotesque in American Literature and Culture* (New York: Columbia University Press, 1997), 75.

27 Cone, *Black Theology,* 23.

28 Srinivas Aravamudan, *Tropicopolitans: Colonialism and Agency, 1688–1804* (Durham, N.C.: Duke University Press, 1999), 269–70, 21. See also Valerie Smith, *Self Discovery and Authority in Afro-American Narrative* (Cambridge, Mass.: Harvard University Press, 1987), 3, and Dana Nelson Salvino, "The Word in Black and White: Ideologies of Race and Literacy in Antebellum America," in *Reading in America: Literature and Social History,* ed. Cathy N. Davidson, 140–56 (Baltimore: Johns Hopkins University Press, 1989).

29 Douglass, *My Bondage,* 106.

30 Ibid., 366.

31 Ibid., 278, 286. Prior to the publication of "The Heroic Slave" in 1853, Douglass's pacifism checked such endorsements of violence. See Larry J. Reynolds, *Righteous Violence: Revolution, Slavery, and the American Renaissance* (Athens: University of Georgia Press, 2011). David Walker's *Appeal* (1828), Henry Highland Garnet's *Address* (1848), and Martin Delany's *Blake* (1859) more consistently suggested that violent self-protection was a viable means for demonstrating the humanity of slaves.

32 Cone, *Black Theology,* 91; Douglass, *My Bondage,* 286.

33 Recall Douglass's well-traveled lecture titled "Self-Made Men" (available at http://memory.loc.gov/). It also bears mentioning that autobiographies of Douglass's era were part of a tradition of "heroic 'American' tales" of personal triumph. Roland L. Williams Jr., *African American Autobiography and the Quest for Freedom* (Westport, Conn.: Greenwood Press, 2000), xii. Douglass's autobiography thus "offers a profound endorsement of the fundamental American plot, the myth of the self-made man." Smith, *Self Discovery,* 27.

34 Gilroy, *Black Atlantic,* 60.

35 James H. Evans Jr., *Spiritual Empowerment in Afro-American Literature* (Lewiston, N.Y.: Edwin Mellen Press, 1987), 3; Henry Louis Gates Jr., ed., introduction to *The Classic Slave Narratives* (New York: Signet Classic, 2002), 2. For more measured accounts of the ways that Douglass both aesthetically "problematize[s]" and politically endorses "the universalizing discourses of human rights," see Paul Gilmore, *Aesthetic Materialism: Electricity and American Romanticism* (Stanford, Calif.: Stanford University Press, 2009), 15, and Lloyd Pratt, "Human beyond Understanding: Frederick Douglass's New Liberal Individual," *Novel* 43, no. 1 (2010): 47–52.

36 Riggins R. Earl Jr., *Dark Symbols, Obscure Signs: God, Self, and the Community in the Slave Mind* (Maryknoll, N.Y.: Orbis Books, 1993), 106. Cassuto argues that Douglass charts a route "from objectification" to "individual empowerment" as a "willful *human* agent." Cassuto, *Inhuman Race,* 27, 86, 87.

37 Johnson, *Persons and Things,* 202; Stephen M. Best, *The Fugitive's Properties: Law and the Poetics of Possession* (Chicago: University of Chicago Press, 2004), 16; Saidiya V. Hartman, *Scenes of Subjection: Terror, Slavery, and Self-Making in Nineteenth-Century America* (New York: Oxford University Press, 1997), 4, 5.

38 See Aimé Césaire, *Discourse on Colonialism,* trans. Joan Pinkham (New York: Monthly Review Press, 2000); Giorgio Agamben, *The Open: Man and Animal* (Stanford, Calif.: Stanford University Press, 2004); and Gayatri Chakravorty Spivak, *A Critique of Postcolonial Reason: Toward a History of the Vanishing Present* (Cambridge, Mass.: Harvard University Press, 1999), esp. 322n15, 324.

39 See Kelly Oliver, *Witnessing: Beyond Recognition* (Minneapolis: University of Minnesota Press, 2001).

40 Douglass, *My Bondage,* 103. The title page attributes the quotation to Coleridge, but it is in fact from the 1845 *Encyclopaedia Metropolitana.* My thanks to Hollis Robbins for bringing this fact to my attention.

41 Douglass, *My Bondage,* 287, 268. Relevant here is James M'Cune Smith's introduction to *My Bondage*: "Naturalists tell us that a full grown man is a resultant . . . of all animated nature on this globe; beginning with the early embryo state, then representing the lowest forms of organic life, and passing through every subordinate grade . . . , until he reaches the last and highest—manhood. In like manner . . . has Frederick Douglass passed through every gradation of rank comprised in our national make-up, and bears upon his person . . . everything that is American"

(132). Referencing the nineteenth-century belief that ontogeny recapitulates phylogeny, the passage nicely illustrates the logic that placed manhood as the highest type of being.

42 Frederick Douglass, "The Claims of the Negro Ethnologically Considered," in *Frederick Douglass: Selected Speeches and Writings,* ed. Philip S. Foner (Chicago: Lawrence Hill, 1999), 284. My reading here resists—but doesn't wholly contradict—that of Ian Frederick Finseth, *Shades of Green: Visions of Nature in the Literature of American Slavery* (Athens: University of Georgia Press, 2009), who argues that *My Bondage* articulates a self-affirming relationship with nature.

43 Douglass, "Claims of the Negro Ethnologically Considered," 284.

44 Douglass, *My Bondage,* 287; Frederick Douglass, "Is It Right to Kill a Kidnapper?," quoted in Reynolds, *Righteous Violence,* 105. See Richard Yarborough, "Race, Violence, and Manhood: The Masculine Ideal in Frederick Douglass's 'The Heroic Slave,'" in *Haunted Bodies: Gender and Southern Texts,* ed. Anne G. Jones and Susan V. Donaldson, 159–84 (Charlottesville: University of Virginia Press, 1997), and Eric A. Goldman, "Spilling Ink and Spilling Blood: Abolitionism, Violence, and Frederick Douglass' *My Bondage and My Freedom,*" *A/B: Auto/Biography Studies* 17, no. 2 (2002): 276–95.

45 Frantz Fanon, *Wretched of the Earth* (New York: Grove Press, 1963), 36–37. To cite only two examples of critics who endorse Douglass's (and Fanon's) logic in this regard: "Violence, not reading, is what set him free" (Cassuto, *Inhuman Race,* 91); "the order of authority on which the slave plantation relied cannot be undone without recourse to the counter-violence of the oppressed" (Gilroy, *Black Atlantic,* 63).

46 Douglass, *My Bondage,* 286, 287; Ralph Waldo Emerson, "Address on the Anniversary of Emancipation of the Negroes in the British West Indies" (Boston: James Munroe, 1844), 31. Reynolds also makes this connection in *Righteous Violence,* 109.

47 Evans, *Spiritual Empowerment,* 3.

48 John Salliant, "Paths Not Taken in North American Abolitionism," *Early American Literature* 46, no. 2 (2011): 383–91.

49 In making this very specific claim, I am of course not suggesting that all slave narratives follow a uniform pattern. For an excellent representation of the diversity of slave narratives, see the University of North Carolina–Chapel Hill's "North American Slave Narratives" website, edited by William L. Andrews (http://docsouth.unc.edu/).

50 Vincent Carretta has recently argued that Equiano actually was born in South Carolina and invented his African origin and Middle Passage

experience as a way to bring attention to his abolitionist cause. Carretta, *Equiano, the African: Biography of a Self-Made Man* (New York: Penguin, 2006). The claim has been hotly contested, but my reading rests less on Equiano's birthplace and more on the fact that his account of traditional Igbo culture largely accords with the findings of recent scholarship. See Finseth, *Shades of Green,* 104; Michael A. Gomez, *Exchanging Our Country Marks: The Transformation of African Identities in the Colonial and Antebellum South* (Chapel Hill: University of North Carolina Press, 1998), 125–27; and Babacar M'Baye, *The Trickster Comes West: Pan-African Influence in Early Black Diasporan Narratives* (Jackson: University Press of Mississippi, 2009).

51 Olaudah Equiano, *The Interesting Narrative of the Life of Olaudah Equiano, or Gustavus Vassa, the African, Written by Himself,* in *Slave Narratives* (New York: Literary Classics of the United States, 2002), 81.

52 Ibid., 86, 200, 203, 60.

53 Following Gates, we could read these scenes as Equiano signifying on the "Western culture" that equates him with "other objects": "When Equiano . . . attempts to speak to the book, there follows only the deafening silence that obtains between two lifeless objects. Only a subject can speak." We should keep in mind, however, that Equiano, like Douglass, rejects not "his master's culture's ultimate sign of subjectivity, the presence of a voice," but only its artificial denial to his race. Gates, *Signifying Monkey,* 156. The distinction is important because, as Aravamudan reminds us, Equiano was a devout Anglican, war veteran, entrepreneur, abolitionist, and, paradoxically, an "occasiona[l] collaborat[or] with the slave trade," a man who sought to ameliorate slavery on the African continent by opening it up to mercantilist trade. Aravamudan, *Tropicopolitans,* 237. See also Baker, *Blues,* 31–51.

54 Equiano, *Interesting Narrative,* 63.

55 For studies that trace Equiano's beliefs to Igbo origins, see Catherine Obianuju Acholonu, *The Igbo Roots of Olaudah Equiano: An Anthropological Approach* (Owerri, Nigeria: Afa, 1989), and *Olaudah Equiano and the Igbo World: History, Society, and Atlantic Diaspora Connections,* ed. Chima J. Korieh (Trenton, N.J.: Africa World, 2009). For things' animateness in ATRs more generally, see Zahan, *Religion, Spirituality, and Thought of Traditional Africa*; Ray, *African Religions*; Evan M. Zuesse, *Ritual Cosmos: The Sanctification of Life in African Religions* (Athens: Ohio University Press, 1979); S. A. Thorpe, *African Traditional Religions: An Introduction* (Pretoria: University of South Africa, 1991); and Theo Sundermeir, *The Individual and Commu-*

nity in African Traditional Religions (Hamburg, Germany: Lit, 1998).

56 Gates, *Signifying Monkey,* 156.

57 "The separation of subject and object—which Western philosophy since Descartes has striven to overcome . . . is not possible in Africa." Sundermeir, *Individual and Community,* 18.

58 See Bibb's recantation of his earlier "great faith in conjuration and witchcraft" in *Narrative of the Life and Adventures of Henry Bibb, an American Slave, Written by Himself,* in *Slave Narratives,* 448; and Brown's less severe but equally dismissive attitude toward African "soothsaying" in *Narrative of William Wells Brown, A Fugitive Slave,* in *Slave Narratives,* 414.

59 Gomez notes that Igbo society did not naturalize slaves' status but allowed them to rise through social ranks. Gomez, *Exchanging Our Country Marks,* 125–27.

60 Césaire, *Discourse on Colonialism,* 42; Stowe, quoted in Johnson, *Persons and Things,* 160.

61 Leonard E. Barrett, *Soul-Force: African Heritage in Afro-American Religion* (Garden City, N.Y.: Anchor Press/Doubleday, 1974), 215, 215–16. See also Kimbwandende Kia Bunseki Fu-Kiau, *African Cosmology of the Bântu–Kôngo, Tying the Spiritual Knot: Principles of Life and Living* (New York: Athelia Henrietta Press, 2001).

62 Stephen Best, "The History of People Who Did Not Exist," paper presented at the "Beyond Critique: Reading after the Hermeneutics of Suspicion" symposium, Duke University, September 10, 2010.

63 Jason R. Young, *Rituals of Resistance: African Atlantic Religion in Kongo and the Lowcountry South in the Era of Slavery* (Baton Rouge: Louisiana State University Press, 2007).

64 "Queer Practices of the Negro Voodoo Doctors," *San Francisco Call,* February 28, 1898, 25. See also "Voodoo Doctors Waxing Wealthy on Foolish Fear," *Washington Times,* June 19, 1904, and "Voodooism in the West Indies," *Los Angeles Herald,* May 14, 1905.

65 For the definitive history of the occupation, see Hans Schmidt, *The United States Occupation of Haiti, 1915–1934* (New Brunswick, N.J.: Rutgers University Press, 1995).

66 Wilbur Forrest, "The Evil Spell of the Voodoo Is over Hayti," *New York Tribune,* December 26, 1920. For other examples of this sensationalized call to arms, see "Voodooism Is the Faith of Haiti," *Jasper News,* January 6, 1921, which includes the descriptive deck "Kill Humans, Drink Blood," and "Haitians Held by Voodooism; Superstitions of African Jungle Prevail on Island; One Priest Killed Children and Drank Their

Blood; Snake Worship Involves Many Extreme Cruelties," *Los Angeles Times,* December 16, 1920.

67 "Cannibals of Hayti, a Fragment of Darkest Africa on the Western Hemisphere," *Salt Lake Herald,* December 27, 1897.

68 Charles Chesnutt, "Superstitions and Folklore of the South" (1901), in *Mother Wit from the Laughing Barrel: Readings in the Interpretation of Afro-American Folklore,* ed. Alan Dundes (Englewood Cliffs, N.J.: Prentice Hall, 1973), 372.

69 For two instances of the turn against the occupation, see "New Deal in Haiti?," *Baltimore Sun,* November 18, 1929, and Major General Smedley Darlington Butler, *War Is a Racket* (New York: Round Table Press, 1935).

70 See Melville J. Herskovits, *Life in a Haitian Valley* (New York: Alfred A. Knopf, 1932); John H. Craige, *Black Bagdad: The Arabian Nights Adventures of a Marine Captain in Haiti* (New York: Minton, Balch, 1933); Richard Loederer, *Voodoo Fire in Haiti* (New York: Doubleday, 1935); and W. B. Seabrook, *The Magic Island* (New York: Harcourt, Brace, 1929).

71 For a classic WPA-funded study of Gullah culture, see the Georgia Writer's Project (Savannah Unit), *Drums and Shadows: Survival Studies among the Georgia Coastal Negroes* (Athens: University of Georgia Press, 1940). See also Zora Neale Hurston, *Tell My Horse: Voodoo and Life in Haiti and Jamaica* (1938), in *Folklore, Memoirs, and Other Writings,* 269–555 (New York: Library of America, 1995); Katherine Dunham, "The Dances of Haiti: Their Social Organization, Classification, Form, and Function," master's thesis, 1939; and Maya Deren, *Divine Horsemen: The Living Gods of Haiti* (New York: Thames and Hudson, 1953).

72 "A White Man's Holiday," *New York Times,* March 15, 1929.

73 Even in the absence of such overt paternalism, voodoo often was reduced to a voyeuristic entertainment, as in the so-called Voodoo *Macbeth* (1936), which was set in Haiti rather than Scotland, performed by the Negro Theater Unit of the WPA Federal Theatre Project, and directed by Orson Welles. See *Weyward Macbeth: Intersections of Race and Performance,* ed. Scott L. Newstok and Ayanna Thompson (New York: Palgrave Macmillan, 2010).

74 Mabel Steedman, *Unknown to the World: Haiti* (London: Hurst and Blackett, 1939), 73–74. For similar rhetoric from the period, see Annice Calland's poem "Voodoo," in *Voodoo* (New York: Harold Vinal, 1926), 1–4, and William J. Makin's travelogue *South of Suez* (London: Jarrolds, 1931), which posits that "civilization has its nights of exhaustion, when

wracked bodies lie in a death-like stupor and the spirit wanders fitful in atavistic dreams" (9).

75 Pauline Hopkins, *Of One Blood: Or, the Hidden Self,* in *The Magazine Novels of Pauline Hopkins* (New York: Oxford University Press, 1988), 558.

76 Ibid., 572, 575, 558, 535.

77 Ibid., 572, 603, 469, 494.

78 Ibid., 576, 448. See William James, "The Hidden Self," *Scribner's Magazine,* March 1890, 361–74.

79 Hopkins, *Of One Blood,* 489, 562.

80 See Thomas J. Otten, "Pauline Hopkins and the Hidden Self of Race," *English Literary History* 59, no. 1 (1992): 227–56; Susan Gillman, "Pauline Hopkins and the Occult: African-American Revisions of Nineteenth-Century Sciences," *American Literary History* 8, no. 1 (1996): 57–82; and Cynthia Schrager, "Pauline Hopkins and William James: The New Psychology and the Politics of Race," in *Female Subjects in Black and White: Race, Feminism, Psychoanalysis,* ed. Elizabeth Abel, Barbara Christian, and Helene Moglen (Berkeley: University of California Press, 1997), 307–29.

81 Mitchell, *Black Belief,* 53; Murphy, *Working the Spirit,* 40.

82 Walker, *Ceremonial Spirit Possession,* 84. See also Keith Cartwright, *Reading Africa into American Literature: Epics, Fables, and Gothic Tales* (Lexington: University of Kentucky Press, 2002); Joseph R. Washington Jr., *Black Sects and Black Cults* (Garden City, N.Y.: Anchor Books, 1973), 1; Sterling Stuckey, *Slave Culture: Nationalist Theory and the Foundations of Black America* (New York: Oxford University Press, 1987), 56; Nicholas C. Cooper and Henry Mitchell, *Soul Theology: The Heart of American Black Culture* (San Francisco: Harper and Row, 1986), x, 5; Gayl Jones, *Liberating Voices: Oral Tradition in African American Literature* (Cambridge, Mass.: Harvard University Press, 1991); Joyce A. Joyce, *Warriors, Conjurers, and Priests: Defining African-Centered Literary Criticism* (Chicago: Third World Press, 1994), 6, 24; and James W. Coleman, *Faithful Vision: Treatments of the Sacred, Spiritual, and Supernatural in Twentieth-Century African American Fiction* (Baton Rouge: Louisiana State University Press, 2005).

83 Hopkins, *Of One Blood,* 468.

84 For the history of the shift from politically minded ex-slave autobiographies to folklore collections, see Bruce Jackson's introduction to *The Negro and His Folklore in Nineteenth-Century Periodicals* (Austin: University of Texas Press, 1967), xv–xxiii.

85 Regarding the African themes and beliefs present in African American

folktales, see Gates, *Signifying Monkey*; Daniel J. Crowly, ed., *African Folklore in the New World* (Austin: University of Texas Press, 1977); Roger D. Abrahams, ed., *Afro-American Folktales: Stories from Black Traditions in the New World* (New York: Pantheon, 1985); William Bascom, *African Folktales in the New World* (Bloomington: Indiana University Press, 1992); and Dickson D. Bruce Jr., *The Origins of African American Literature, 1680–1865* (Charlottesville: University Press of Virginia, 2001).

86 See Michael A. Elliott, *The Culture Concept: Writing and Difference in the Age of Realism* (Minneapolis: University of Minnesota Press, 2002); Brad Evans, *Before Cultures: The Ethnographic Imagination in American Literature, 1865–1920* (Chicago: University of Chicago Press, 2005); and Bill Brown, *A Sense of Things: The Object Matter of American Literature* (Chicago: University of Chicago Press, 2003).

87 In *What Is Posthumanism?* (Minneapolis: University of Minnesota Press, 2010), Cary Wolfe argues that posthumanism must entail the troubling of both humanist form (the assumption that there is a natural, one-to-one correspondence between the representation and the represented) and humanist content (the taxonomic privileging of the human over the nonhuman).

88 For the relation of these movements and Africanist beliefs, see Anderson, *Conjure*, 140–42.

89 Although this conception owes much to British romanticism, both Muir and Burroughs were inspired more directly by Emerson, Thoreau, and Whitman. Indeed, Burroughs wrote the first biography on Whitman, *Notes on Walt Whitman as Poet and Person* (1867). In the later *Birds and Poets* (the title resonating with Whitmanian echoes), he characteristically claimed that "the true poet knows more about Nature than the naturalists because he carries her open secrets in his heart." http://wakerobin.org/.

90 H. P. Blavatsky, *The Secret Doctrine: The Synthesis of Science, Religion, and Philosophy*, vol. I, *Cosmogenesis* (1888; repr., Pasadena, Calif.: Theosophical University Press, 1952), 268, 445. See also Richard Maurice Bucke's *Cosmic Consciousness: A Study in the Evolution of the Human Mind* (1901; repr., New York: E. P. Dutton, 1967), which, quoting from Dante, contends that the cosmically conscious person is "'*transhumanized* into a God'" (75; my emphasis).

91 Mechal Sobel, *Trabelin' On: The Slave Journey to an Afro-Baptist Faith* (Westport, Conn.: Greenwood Press, 1979), 10.

92 Levi Bryant, *The Democracy of Objects* (Ann Arbor, Mich.: Open Humanities Press, 2011); Bruno Latour, *We Have Never Been Mod-*

ern, trans. Catherine Porter (Cambridge, Mass.: Harvard University Press, 1993), 142; and Isabelle Stengers, *Cosmopolitics I,* trans. Robert Bononno (Minneapolis: University of Minnesota Press, 2010).

93 See Barbara Christian, “The Race for Theory,” *Cultural Critique* 6 (Spring 1987): 51–63, and bell hooks, “Postmodern Blackness,” in *The Norton Anthology of Theory and Criticism,* ed. Vincent B. Leitch (New York: W. W. Norton, 2001), 2478–84.

94 For contemporary neoanimisms, see Bruno Latour, *Politics of Nature: How to Bring the Sciences into Democracy,* trans. Catherine Porter (Cambridge, Mass.: Harvard University Press, 2004), 224: “Nothing is more anthropocentric than the inanimism of nature”; also Timothy Morton, *The Ecological Thought* (Cambridge, Mass.: Harvard University Press, 2010), 8, and Jane Bennett, *Vibrant Matter: A Political Ecology of Things* (Durham, N.C.: Duke University Press, 2010).

4. HODOO YOU THINK YOU ARE?

1 Charles Chesnutt, “Superstitions and Folklore of the South,” *Modern Culture* 13 (1901): 231–35, repr. in *Mother Wit from the Laughing Barrel: Readings in the Interpretation of Afro-American Folklore,* ed. Alan Dundes (Englewood Cliffs, N.J.: Prentice Hall, 1973), 372.

2 Ibid., 371, 370, 373. See also Tyrie J. Smith, “‘Your People Will Never Rise in the World’: Chesnutt’s Message to a Black Readership,” in *Charles Chesnutt Reappraised: Essays on the First Major African American Fiction Writer,* ed. David Garrett Izzo and Maria Orban, 110–19 (Jefferson, N.C.: McFarland, 2009).

3 Eric J. Sundquist, *To Wake the Nations: Race in the Making of American Literature* (Cambridge, Mass.: Belknap Press of Harvard University Press, 1993), 295, 296.

4 Richard H. Brodhead, ed., introduction to *The Conjure Woman and Other Conjure Tales* by Charles Chesnutt (Durham, N.C.: Duke University Press, 1993), 13.

5 Chesnutt was thus representative of what Curtis J. Evans has identified as postwar “Northern middle-class black[s’] . . . ambivalen[ce] about the folk culture of Southern blacks. . . . They rejected whites’ racial interpretations of black religion as African regression . . . [but] also found much that was distasteful and culturally inferior in” it. Evans, *The Burden of Black Religion* (New York: Oxford University Press, 2008), 12–13.

6 Jeffrey E. Anderson, *Conjure in African American Society* (Baton Rouge: Louisiana State University Press, 2005), 4.

7 Louis Pendleton, “Notes on Negro Folk-Lore and Witchcraft in the

South," *Journal of American Folklore* 3, no. 10 (1890): 204. A later article expresses similar incredulity at the continued survival of "the superstitions and beliefs of the negro race": "These beliefs were brought here from Africa by the first comers and continue in full force to this day, notwithstanding the negro is a freeman and living amongst the white people of the United States of America, who are probably as practical as any human beings on earth." Julien A. Hall, "Negro Conjuring and Tricking," *Journal of American Folklore* 10, no. 38 (1897), 241. Formed in 1888, the *Journal of American Folklore* devoted numerous pages—some bemused, some judgmental—to documenting turn-of-the-century examples of conjuration. See also "Concerning Negro Sorcery in the United States," *Journal of American Folklore* 3, no. 11 (1890): 281–87, and "Conjuring and Conjure-Doctors in the Southern United States," *Journal of American Folklore* 9, no. 33 (1896): 143–47.

8 Chesnutt, "Superstitions," 370. For more details on Chesnutt's biography, especially his early years in North Carolina, see Helen M. Chesnutt, *Charles Waddell Chesnutt: Pioneer of the Color Line* (Chapel Hill: University of North Carolina Press, 1952), and Ernestine Williams Pickens, *Charles W. Chesnutt and the Progressive Movement* (New York: Pace University Press, 1994).

9 Charles Chesnutt, quoted in J. N. Heermance, *Charles W. Chesnutt: America's First Great Black Novelist* (Hamden, Conn.: Archon, 1974), 69. According to Anderson, a high percentage of African Americans in the South, as well as a significant number of whites, practiced conjure both before and after emancipation, its influence looming large in social, economic, and political spheres. For the most complete records of conjure practices and beliefs in the United States, see Newbell Niles Puckett, *Folk Beliefs of the Southern Negro* (1926; repr., Negro Universities Press, 1968), and Harry Middleton Hyatt, *Hoodoo, Conjuration, Witchcraft, Rootwork: Beliefs Accepted by Many Negroes and White Persons, These Being Orally Recorded among Blacks and Whites,* 5 vols. (Hannibal, Mo.: Western, 1970).

10 For scholars who rightly have identified this criticism in the conjure tales, see, among others, William Andrews, *The Literary Career of Charles W. Chesnutt* (Baton Rouge: Louisiana State University Press, 1980); Lorne Fienberg, "Charles W. Chesnutt and Uncle Julius: Black Storytellers at the Crossroads," *Studies in American Fiction* 15, no. 2 (1988): 161–74; Craig Werner, "The Framing of Charles W. Chesnutt: Practical Deconstruction in the Afro-American Tradition," in *Southern Literature and Literary Theory,* ed. Jefferson Humphries (Athens:

University of Georgia Press, 1990), 239–52; Ben Slote, "Listening to 'The Goophered Grapevine' and Hearing Raisins Sing," *American Literary History* 6, no. 4 (1994): 684–94; Robert C. Nowatzi, "'Passing' in a White Genre: Charles W. Chesnutt's Negotiations of the Plantation Tradition in *The Conjure Woman*," *American Literary Realism 1870–1910* 27, no. 2 (1995): 20–36; Stephen P. Knadler, "Untragic Mulatto: Charles Chesnutt and the Discourse of Whiteness," *American Literary History* 8, no. 3 (1996): 426–48; and Henry B. Wonham, *Charles W. Chesnutt: A Study of the Short Fiction* (New York: Twayne, 1998).

11 Charles Chesnutt, "The Conjurer's Revenge," in *Conjure Woman and Other Conjure Tales,* 79; Chesnutt, "Sis' Becky's Pickaninny," in *Conjure Woman and Other Conjure Tales,* 83. The specifications for the rabbit foot match those reported in "Concerning Negro Sorcery in the United States," 283.

12 Chesnutt, "Sis' Becky's Pickaninny," 83.

13 Brodhead, introduction, 14.

14 Sundquist, *To Wake the Nations,* 358, 358–59.

15 Ibid., 359–60. "Why the Darkey Loves Chicken" is the subtitle of "A Victim of Heredity," in *Conjure Woman and Other Conjure Tales,* 172–82.

16 Sundquist, *To Wake the Nations,* 365, 359.

17 Ibid., 385.

18 For earlier examples of the conjure–signifying and Chesnutt–Julius–Peggy conjunctions, see Melvin Dixon, "The Teller as Folk Trickster in Chesnutt's *The Conjure Woman,*" *College Language Association Journal* 18 (1974): 186–97; Marjorie Pryse, "Introduction: Zora Neale Hurston, Alice Walker, and the 'Ancient Power' of Black Women," in *Conjuring: Black Women, Fiction, and Literary Tradition,* ed. Marjorie Pryse and Hortense Spillers, 1–24 (Bloomington: Indiana University Press, 1985); and Houston Baker Jr., *Modernism and the Harlem Renaissance* (Chicago: University of Chicago Press, 1987), 46.

19 Matthew R. Martin, "The Two-Faced New South: The Plantation Tales of Thomas Nelson Page and Charles W. Chesnutt," *Southern Literary Journal* 30, no. 2 (1998): 32.

20 Charles Chesnutt, *The Journals of Charles W. Chesnutt,* ed. Richard H. Brodhead (Durham, N.C.: Duke University Press, 1993), 154. Given his adamant *de*-emphasis of racial distinctions (in rejecting "race pride" as "a modern invention of the white people," he once asked, "Why and to what good end, should we wish to perpetuate this disastrous difference?" ["Race Prejudice: Its Causes and Its Cures" (1905), in *Charles W.*

Chesnutt: Essays and Speeches, ed. Joseph R. McElrath Jr., Robert C. Leitz III, and Jesse S. Crisler (Stanford, Calif.: Stanford University Press, 1999), 232]), it's questionable whether Chesnutt wanted to "merge . . . his identity with that of other Black people." Joyce A. Joyce, *Warriors, Conjurers, and Priests: Defining African-Centered Literary Criticism* (Chicago: Third World Press, 1994), 39–40.

21 See Charles Chesnutt, "The Future American" (1900), in *Charles W. Chesnutt: Essays and Speeches,* 121–36. On Chesnutt's combatting of racism through the deconstruction, not transvaluation, of race, see Dean McWilliams, *Charles W. Chesnutt and the Fictions of Race* (Athens: University of Georgia Press, 2002); Heermance, *Charles W. Chesnutt,* 37–41; Francis R. Keller, *An American Crusade* (Salt Lake City, Utah: Brigham Young University Press, 1978), 49; and Sylvia Render, *Charles W. Chesnutt* (Boston: Twayne, 1980), 92.

22 In the African diaspora, a trickster can be a conjurer (and vice versa), but there is no necessary identity between the two. See Gates, *Signifying Monkey,* 6. One might say it's the difference between Peggy and Julius—a difference most Chesnutt critics fail to honor, making Julius the central "conjurer" of *The Conjure Woman.*

23 Sundquist, *To Wake the Nations,* 358, 370, 372.

24 Ibid., 383. On this aspect of Chesnutt's later fiction, see McWilliams, *Charles W. Chesnutt,* and Matthew Wilson, *Whiteness in the Novels of Charles W. Chesnutt* (Jackson: University of Mississippi Press, 2004). If the conjure tales are read only as a watered-down version of the more overtly political novels, however, then we are faced with the paradox that Joyce unintentionally reveals: "Chesnutt's conjuring powers were so effective that white readers failed to see . . . the focus of Chesnutt's criticism. . . . The general opinion at the time of publication was, in fact, that the author of *Conjure Woman* was white." Joyce, *Warriors, Conjurers, and Priests,* 41. Here success is marked by a *failure* of recognition, ostensibly the goal of Chesnutt's critique.

25 For two Chesnutt novels that reference lynching, see *The Marrow of Tradition* (New York: Penguin, 1993) and *Mandy Oxendine* (Urbana: University of Illinois Press, 1997). See also Sandra Gunning's chapter on these texts in *Race, Rape, and Lynching: The Red Record of American Literature,* 48–76 (New York: Oxford University Press, 1996) and John Mac Kilgore, "The Cakewalk of Capitalism in Charles Chesnutt's *The Marrow of Tradition,*" *American Literature* 84, no. 1 (2012): 61–87. For a commentary by Ida B. Wells, Chesnutt's contemporary, see "Southern Horrors: Lynch Law in All Its Phases" (1892), in *Southern Horrors and*

Other Writings: The Anti-Lynching Campaign of Ida B. Wells, 1892–1900, ed. Jacqueline Jones Royster, 49–72 (Boston: Bedford/St. Martin's, 1997). More generally, see Stewart E. Tolnay and E. M. Beck, *A Festival of Violence: An Analysis of Southern Lynchings, 1882–1930* (Urbana: University of Illinois Press, 1995), and Amy Louise Wood, *Lynching and Spectacle: Witnessing Racial Violence in America, 1890–1940* (Chapel Hill: University of North Carolina Press, 2009).

26 Paul Outka, *Race and Nature from Transcendentalism to the Harlem Renaissance* (New York: Palgrave Macmillan, 2008), 160.

27 John Muir, *Our National Parks* (1901), in *The Eight Wilderness-Discovery Books* (London: Diadem Books, 1992), 459; quoted in slightly different form in Outka, *Race and Nature*, 156.

28 Outka, *Race and Nature*, 104. See also Ian Frederick Finseth, *Shades of Green: Visions of Nature in the Literature of American Slavery, 1770–1860* (Athens: University of Georgia Press, 2009).

29 See Roger D. Abrahams's introduction to *Afro-American Folktales: Stories from Black Traditions in the New World*, 3–36 (New York: Pantheon, 1985), for the amorality of conjure practice.

30 The lone exception to conjure's absence in the frame narratives may be Julius's rabbit's foot in "Sis' Becky's Pickaninny," but this is obviously of a different order than Peggy's conjure.

31 On the mutual influence of local color fiction with early ethnography, see Michael A. Elliott, *The Culture Concept: Writing and Difference in the Age of Realism* (Minneapolis: University of Minnesota Press, 2002).

32 For Chesnutt's claims to his original authorship of *The Conjure Woman* tales, see his "Post-Bellum—Pre-Harlem" (1931), in *Charles W. Chesnutt: Stories, Novels, and Essays*, 906–12 (New York: Library of America, 2002). Thirty years prior, in "Superstitions and Folklore of the South," however, Chesnutt had admitted that, although "I took, or thought I did, considerable liberty with my subject," he eventually "discovered that the brilliant touches, due, I had thought, to my own imagination, were after all but dormant ideas, lodged in my mind by old Aunt This and old Uncle That" (372). For more on this constellation of claims, see Dundes's headnote to Chesnutt's article in *Mother Wit from the Laughing Barrel*; Andrews, *Literary Career*, 48; Sundquist, *To Wake the Nations*, 296–98; and Slote, "Listening to 'The Goophered Grapevine,'" 689.

33 In this regard, it is telling that conjurers in general and Peggy in particular are depicted as living on the outskirts of human communities rather than within them.

34 The criticism has verified this claim (while noting some variations in Chesnutt's version) as well as expanded it by showing the large extent to which all of Chesnutt's conjure tales rely on diasporic folk traditions. See Robert Hemenway, "The Functions of Folklore in Charles Chesnutt's *The Conjure Woman*," *Journal of the Folklore Institute* 13 (1976): 283–309; Gloria C. Oden, "Chesnutt's Conjure as African Survival," *MELUS* 5, no. 1 (1978): 38–48; Elon A. Kulii, "Poetic License and Chesnutt's Use of Folklore," *CLA Journal* 38, no. 2 (1994): 247–53; and Tiel Lundy, "With Myriad Subtleties: Recognizing an Africanist Presence in Charles W. Chesnutt's *The Conjure Woman*," in Izzo and Orban, *Charles Chesnutt Reappraised*, 173–87.

35 Charles Chesnutt, "The Goophered Grapevine," in *The Conjure Woman and Other Conjure Tales*, 31, 35; hereinafter cited parenthetically by page number.

36 As Oden notes, "the use of conjure to protect property from theft is one of the thirty-five basic situations for which West Africans use conjure." Oden, "Chesnutt's Conjure," 41.

37 For a good example of this line of argumentation, see Wonham, *Charles W. Chesnutt*, 15, 17. I emphasize the use of the term *doctor* here because conjurers are often colloquially referred to as either "root doctors" or "two-headed doctors." Moreover, the trope of conjure assisting people after "white" medicine fails—and the related idea of the latter being helpless to cure conjure-incurred harm—is pervasive in the popular tradition. See, for instance, the unpublished recording AFS 14268a included in "Root Doctor: A Folklore Study in Which a Negro Healer Explains His Work before a Concealed Microphone to Harry M. Hyatt," part of the Harry M. Hyatt Collection, Library of Congress, Washington, D.C. Made in the late 1930s or early 1940s, the recording gives an example of a "root doctor" curing a woman after white doctors couldn't help remove the "frogs and lizards" and "things running around in [her] body."

38 Jeffrey Myers, "Other Nature: Resistance to Ecological Hegemony in Charles W. Chesnutt's *The Conjure Woman*," *African American Review* 37, no. 1 (2003): 10. See also Myers, *Converging Stories: Race, Ecology, and Environmental Justice in American Literature* (Athens: University of Georgia Press, 2005), 87–110, and Lee Rozelle, *Ecosublime: Environmental Awe and Terror from New World to Oddworld* (Tuscaloosa: University of Alabama Press, 2006).

39 Myers argues that McAdoo's (and, by extension, John's) "subjectivity . . . depends on a separation of the self from the environment, an illusory

separation that Julius's tale deconstructs." Myers, "Other Nature," 10. "Such a self," Myers continues, "lives in constant fear that its bodily sameness with the physical world and with other human bodies will be exposed.... To maintain the illusion of separateness and superiority, the Euroamerican self must constantly display his mastery over the material world." Myers, *Converging Stories*, 16. See also Ellen J. Goldner, "Other(ed) Ghosts: Gothicism and the Bonds of Reason in Melville, Chesnutt, and Morrison," *MELUS* 24, no. 1 (1999): 59–83, esp. 60–61.

40 Rozelle, *Ecosublime*, 24.

41 Other Chesnutt stories make explicit the fact that slave masters, and by extension all white people, do not escape the rules of the conjure universe. "Mars Jeems's Nightmare," for example, tells of a harsh slave owner being transformed by conjure into a slave himself, after which he is beaten by his own overseer and made to work his own land. Part of the transformation's poignancy, of course, is its subversion of the racist assumption that there is an unbridgeable ontological divide between white "persons" and nonwhite "things."

42 Myers, "Other Nature," 6.

43 Ibid., 11, 13.

44 Ibid., 6.

45 Rozelle, *Ecosublime*, 28. Rozelle emphasizes that conjure defies the self-serving "idea of a natural immersion so touted by British romantics and American transcendentalism" (23), but at times his account appears to parallel it.

46 Myers, "Other Nature," 8; Rozelle, *Ecosublime*, 24, 26; Myers, "Other Nature," 10.

47 Wonham argues that such transformations make their African American subjects "vulnerable to new forms of exploitation." Wonham, *Charles W. Chesnutt*, 20. However, the forms he has in mind are economic rather than cosmological.

48 Myers, "Other Nature," 6; Rozelle, *Ecosublime*, 23.

49 On the period's environmental devastations and the responses they provoked, see *Major Problems in American Environmental History*, ed. Carolyn Merchant, 2nd ed. (Boston: Houghton-Mifflin, 1993); Theodore Steinberg, *Nature Incorporated: Industrialization and the Waters of New England* (Amherst: University of Massachusetts Press, 1994); and David Mazel, *American Literary Environmentalism* (Athens: University of Georgia Press, 2000).

50 For the shift in logging, see Michael Williams, *Americans and Their*

Forests: A Historical Geography (New York: Cambridge University Press, 1989).

51 John Muir, *The Yosemite* (1912), in *Eight Wilderness-Discovery Books,* 714.

52 John Muir, *My First Summer in the Sierra* (1911), in *Eight Wilderness-Discovery Books,* 235. Muir had used the same line almost verbatim seventeen years earlier (including, however, a reference to God that offsets the animism or pantheism of the later version): "every pulse beats high, every life-cell rejoices, the very rocks seem to tingle with life, and God is felt brooding over everything great and small." Muir, *The Mountains of California* (1894), in *Eight Wilderness-Discovery Books,* 368. Muir's contemporary, John Burroughs, offers a similar if slightly less theological picture of our self-affirming relationship with the natural world: "I find something akin to poetry and religion . . . in the shows of day and night, and in my excursions to fields and woods." Burroughs, *The Breath of Life* (Boston: Houghton Mifflin, 1915), vii.

53 Outka, *Race and Nature,* 104.

54 Timothy Morton, *The Ecological Thought* (Cambridge, Mass.: Harvard University Press, 2010), 15, 16.

55 Ibid., 7, 16. See also Morton, *Ecology without Nature* (Cambridge, Mass.: Harvard University Press, 2007); Morton, "Queer Ecology," *PMLA* 125, no. 2 (2010): 273–82; and Dana Phillips's *The Truth of Ecology: Nature, Culture, and Literature in America* (New York: Oxford University Press, 2003), which diagnoses the imprecisions and contradictions of much "devotional" ecocriticism (ix). For a useful collection of primary and secondary reflections on the kinds of romantic discourses on nature to which Morton and Nelson are responding, see "Part I: Romantic Ecology and Its Legacy," in *The Green Studies Reader: From Romanticism to Ecocriticism,* ed. Laurence Coupe, 13–58 (New York: Routledge, 2000).

56 Rozelle, *Ecosublime,* 22. Rozelle here draws on the work of Christopher Hitt, "Toward an Ecological Sublime," *New Literary History* 30, no. 3 (1999): 603–23. Hitt attempts to recuperate the sublime from an "ideology [that] is fundamentally or intrinsically maleficent" by slowing the all-too-quick move from "humbling fear" to "ennobling validation for the perceiving subject," which "transform[s]" "humility" into "self-apotheosis" and "validat[es] the individual's dominion over the nonhuman world" (605, 606, 608). Despite wanting to "resist [this] traditional reinscription of humankind's supremacy over nature" (609), Hitt still insists that the sublime is about "an individual's process

of self-discovery" via a "confront[ation] with the wild otherness of nature" (610, 620). This insistence seems inconsistent with Hitt's claim elsewhere in the essay that "there can be no subject or object, since this very dualism is a conceptual construction" (614).

57 Outka, *Race and Nature,* 14, 15.

58 Morton, *Ecological Thought,* 15.

59 Ibid.

60 Bruce Clarke, *Posthuman Metamorphosis: Narrative and Systems* (New York: Fordham University Press, 2008), 164.

61 The Library of Congress has a recording of a mid-twentieth-century African American folk tale, "The Haunted Pine Tree" (AFS 8795A3; spoken by L. Crump, recorded in Jonas Ridge, August 1940) that shares some formal similarities with Julius's tale. See also recording 14268B in the Harry M. Hyatt Collection.

62 Myers, "Other Nature," 5. See also Rozelle, *Ecosublime,* 27–30. On such transformations being based in diasporic spiritual beliefs, see Oden, "Chesnutt's Conjure," 42, and Puckett, *Folk Beliefs,* 160–62, 205.

63 Charles Chesnutt, "Po' Sandy," in *Conjure Woman and Other Conjure Tales,* 50, 51; hereinafter cited parenthetically by page number. The horror of the scene is heightened by Julius's dwelling on the fact that the tree itself resists being cut, "creakin,' an shakin,' en wobbling,'" "sweekin,' en moanin,' en groanin,'" while first the axes and then the mill-saw labor to cut through the strangely tough wood ("Po' Sandy," 50, 51).

64 Sundquist, *To Wake the Nations,* 376; Myers, "Other Nature," 5. See also Fienberg, "Charles W. Chesnutt"; Outka, *Race and Nature*; and Charles Duncan, *The Absent Man: The Narrative Craft of Charles W. Chesnutt* (Athens: Ohio University Press, 1998).

65 For the Africanist roots of this metamorphosis, see Oden, "Chesnutt's Conjure," 42. Bill Brown argues that the tale demonstrates "the power of animism (actual or fabricated) to interrupt the Northern appropriation of Southern property . . . [and] retard the course of change and 'progress.'" Brown, *A Sense of Things: The Object Matter of American Literature* (Chicago: University of Chicago Press, 2003), 117–18. I would agree except to note that it is only the frame, not the tale itself, that hedges regarding whether animism is "actual or fabricated."

66 Muir, *First Summer in the Sierra,* 251.

67 Myers imputes the macabre ending of "Po' Sandy" to the white characters' "exploitation of the land in the form of logging," claiming that their profit-driven domination of the environment upsets the organic

"[r]egard for [the] landscape," the "intimate relationship with it," that characterizes African Americans' "exist[ing] in harmony with the landscape" ("Other Nature," 5, 12, 13). However, other tales of transformation also end badly without featuring whites' ecological destruction. "Tobe's Tribulations," for instance, follows the title character as he attempts to escape his enslavement by being made into various creatures, including, finally, a frog. Through an accident, however, he is never changed back, meaning that he remains in frog form even after emancipation: "It's be'n fifty yeahs er mo.' . . . But he ain' nebber got ti'ed er wantin' ter be tu'nt back ter hisse'f. . . . Co'se ef he had waited lak de res' un us he'd a be'n free long ago; but he didn' know dat, en he doan knows it yet." Charles Chesnutt, "Tobe's Tribulations," in *Conjure Woman and Other Conjure Tales,* 192.

68 Myers, "Other Nature," 16.

69 Clay eating is believed to have originated in Africa and then been brought to the South by slaves, where it was used as a dietary supplement, a conjure practice, and a means to escape work by inducing sickness or death. The practice of geophagy continues to this day among both white and black populations in the South, with some scientists claiming that the ingested clay provides minerals otherwise unavailable in the consumers' diets. See J. Montgomery M'Govern, "The Clay Eaters of Carolina," in *Everybody's Magazine, Vols. I and II,* 246–49 (New York: North American, 1900); "Clay Eaters," *New York Times,* May 15, 1904; Tom Painter and Roger Kammerer, "The Clay-Eaters of Edgecombe County," in *Forgotten Tales of North Carolina* (Charleston, S.C.: History Press, 2006), 59; and Jennifer L. Fleissner's superb "Earth-Eating, Addiction, Nostalgia: Charles Chesnutt's Diasporic Regionalism," *Studies in Romanticism* 49, no. 2 (2010): 313–36.

70 Charles Chesnutt, "Lonesome Ben," in *Conjure Woman and Other Conjure Tales,* 155.

71 Ibid., 156.

72 Ibid.

73 Charles Chesnutt, "The Gray Wolf's Ha'nt," in *Conjure Woman and Other Conjure Tales,* 95, 94, 95. The quotation appears to be one of Herbert Spencer's principles of synthetic or universal evolution. See Spencer, *First Principles of a New System of Philosophy,* 2nd ed. (New York: D. Appleton, 1870), 285. See also McWilliams's discussion of this and other passages as possibly engaging Spencer's racism in *Charles W. Chesnutt,* 91. Baker rightly claims that the story holds the tedium and assumed epistemological superiority of John's "study of philosophy"

(*Modernism,* 95) up to ridicule, but I disagree with what Baker takes to be the correlative of this claim: that Julius's tale offers a more self- and race-affirming cosmology than John's (44–46); it is more vivid and corporeal, yes, but not more affirmative (one need only consider the annihilation of Ben).

74 Charles Chesnutt, "Mars Jeems's Nightmare," in *Conjure Woman and Other Conjure Tales,* 55.

75 Chesnutt, "Tobe's Tribulations," 185.

76 Chesnutt, "Mars Jeems's Nightmare," 55.

77 On the history of this worldview, see Arthur O. Lovejoy, *The Great Chain of Being: A Study of the History of an Idea* (Cambridge, Mass.: Harvard University Press, 1976).

78 For an early instance, see Robert Stepto, "The Simple but Intensely Human Inner Life of Slavery: Storytelling, Fiction, and the Revision of History in Charles W. Chesnutt's 'Uncle Julius Stories,'" in *History and Tradition in Afro-American Culture,* ed. Gunter H. Lenz, 29–55 (Frankfurt, Germany: Campus, 1984).

79 Chesnutt, "Tobe's Tribulations," 185; Charles Chesnutt, "Dave's Neckliss," in *Conjure Woman and Other Conjure Tales,* 124.

80 Chesnutt, "Tobe's Tribulations," 192; Chesnutt, "Conjurer's Revenge," 71. Compare with Muir, *First Summer in the Sierra,* 248: "When we try to pick out anything by itself, we find it hitched to everything else in the universe. One fancies a heart like our own must be beating in every crystal and cell, and we feel like stopping to speak to the plants and animals as friendly fellow mountaineers." Although Muir's vision and the conjure cosmology share a sense of interconnectivity (nothing is isolatable because everything is "hitched to everything else"), the former presents it in the rhetoric of empathetic friendship, whereas the latter stipulates only a common vulnerability to harm.

81 Morton, *Ecological Thought,* 8.

82 See James Haskins, *Voodoo and Hoodoo: Their Tradition and Craft as Revealed by Actual Practitioners* (New York: Stein and Day, 1978), 29; Henry Mitchell, *Black Belief: Folk Beliefs in America and West Africa* (New York: Harper and Row, 1975), 72; Robert Farris Thompson, *Flash of the Spirit: African and Afro-American Art and Philosophy* (New York: Random House, 1983), 117–45; and Will Coleman, *Tribal Talk: Black Theology, Hermeneutics, and African/American Ways of "Telling the Story"* (University Park: Pennsylvania State University Press, 2000), 13.

83 See Jacques Derrida, *The Beast and the Sovereign, Vol. I,* trans. Geoffrey Bennington (Chicago: University of Chicago Press, 2009). Following

Derrida, Morton, *Ecological Thought,* 8, argues that "being a person means never being sure that you're one. In an age of ecology without Nature, we would treat many more beings as people while deconstructing our ideas about what counts as people."

84 According to Haskins, in the conjure cosmos, the human "sees [itself] as an integral part of the universe. There is no . . . sense of separation or differentiation from surrounding forces. . . . Indeed, [its] very . . . being . . . is defined in terms of the people and the world around [it]." Haskins, *Voodoo and Hoodoo,* 27–28. For a cognate belief in ATRs, see Dominique Zahan, *The Religion, Spirituality, and Thought of Traditional Africa,* trans. Kate Ezra Martin and Lawrence M. Martin (Chicago: University of Chicago Press, 1979), 9: "the individual does not constitute a closed system in opposition to the outside world in order to better secure his own substance and limitations. On the contrary, he enters into the surrounding environment, which in turn penetrates him."

85 Fleissner, in "Earth-Eating," 315, makes a related point when she links Julius's remembrances of slavery, figured in scenes of his and other characters' eating, to Derrida's *pharmakon,* "at once a painful symptom and a creative resource."

86 Emmanuel Levinas, *Ethics and Infinity: Conversations with Philippe Nemo,* trans. Richard A. Cohen (Pittsburgh, Pa.: Duquesne University Press, 1985), 95. See also Levinas, *Otherwise than Being, or, Beyond Essence,* trans. Alphonso Lingis (Pittsburgh, Pa.: Duquesne University Press, 1998).

87 On this essential circumscription or limit in conjure practice, see Puckett, *Folk Beliefs,* 134, and Haskins, *Voodoo and Hoodoo,* 78.

5. "IT MIGHT BE THE DEATH OF YOU"

1 As discussed in chapter 2, pre-Boasian anthropology and ethnology were typified largely by biological and/or metaphysical narratives of absolute racial difference. Boas popularized the alternative notion that racial differences were attributable to historically mutable cultures rather than static, natural hierarchies, an idea that became central in the development of anthropology as an academic and professional field. See Marvin Harris, *The Rise of Anthropological Theory: A History of Theories of Culture,* rev. ed. (Walnut Creek, Calif.: AltaMira Press, 2001).

2 Zora Neale Hurston, *Mules and Men* (1935; repr., New York: Harper and Row, 1990), 251, 273; hereinafter cited parenthetically by page number. Boas's preface praises Hurston's status as a "southern Negro"

(she was "one of them and was fully accepted as such by the companions of her childhood"), which afforded her the opportunity to "penetrate through the affected demeanor by which the Negro excludes the White observer effectively from participating in his true inner life" (*Mules and Men,* xiii). She therefore offers a unique contribution to "our knowledge" of the subject (*Mules and Men,* xiii). As these quotations indicate, Boas leaves ambiguous whether Hurston is included in—or is simply an object of—"our knowledge." For more on the Hurston–Boas relationship, see Michael A. Elliot, *The Culture Concept: Writing and Difference in the Age of Realism* (Minneapolis: University of Minnesota Press, 2002), 161–88.

3 See, for instance, Alice Walker's praise of the "authenticity of her material . . .verified by her familiarity with its context" in the foreword to *Zora Neale Hurston: A Literary Biography,* by Robert E. Hemenway (Urbana: University of Illinois Press, 1977), xii, and Hazel V. Carby, "The Politics of Fiction, Anthropology, and the Folk: Zora Neale Hurston," in *Cultures in Babylon: Black Britain and African America,* 168–88 (New York: Verso, 1999).

4 In an August 20, 1934, letter to Boas, Hurston explained that much of the book's "unscientific matter," such as the "between-story conversation and business" that foregrounded Hurston, was present because her publisher "want[ed] a very readable book that the average reader [could] understand." The Franz Boas Papers, Library of Congress, Washington, D.C. Many scholars, however, have not been convinced that this accounts for Hurston's prominence in *Mules and Men.* D. A. Boxwell notes that Hurston's "presentation of herself as a strong, ordering force in the text" is "unique to the ethnological writing of her time." "'Sis Cat' as Ethnographer: Self-Presentation and Self-Inscription in Zora Neale Hurston's *Mules and Men,*" *African American Review* 26, no. 4 (1992): 610. Other critics have read an aesthetic, epistemological, and/or ideological choice into it. See David Headon, "'Beginning to See Things Really': The Politics of Zora Neale Hurston," in *Zora in Florida,* ed. Steve Glassman and Kathryn Lee Seidel, 28–37 (Orlando: University of Central Florida Press, 1991); Brian Carr and Tova Cooper, "Zora Neale Hurston and Modernism at the Critical Limit," *Modern Fiction Studies* 48, no. 2 (2002): 285–313; Lori Jirousek, "'That Commonality of Feeling': Hurston, Hybridity, and Ethnography," *African American Review* 38, no. 3 (2004): 417–27; and Lynda Hoffman-Jeep, "Creating Ethnography: Zora Neale Hurston and Lydia Cabrera," *African American Review* 39, no. 3 (2005): 337–53.

5 Ayana I. Karanja, *Zora Neale Hurston: The Breath of Her Voice* (New

York: Peter Lang, 1999), 6. See also Elizabeth Jane Harrison, "Zora Neale Hurston and Mary Hunter Austin's Ethnographic Fiction: New Modernist Narratives," in *Unmanning Modernism: Gendered Re-readings,* ed. Elizabeth Jane Harrison and Shirley Peterson, 44–58 (Knoxville: University of Tennessee Press, 1997), esp. 46; David C. Estes, "The Neo-African Vatican: Zora Neale Hurston's New Orleans," in *Literary New Orleans in the Modern World,* ed. Richard S. Kennedy, 66–82 (Baton Rouge: Louisiana State University Press, 1998), esp. 69; and Sandra Dolby-Stahl, "Literary Objective: Hurston's Use of Personal Narrative in *Mules and Men,*" in *Critical Essays on Zora Neale Hurston,* ed. Gloria L. Cronin, 43–52 (New York: G. K. Hall, 1998).

6 John Dorst, "Rereading *Mules and Men*: Toward the Death of the Ethnographer," *Cultural Anthropology* 2, no. 3 (1987): 313. In one such story, Ole Massa, dressed in a bedsheet, tricks his slave, John, into believing that he is the Lord, subsequently trying to persuade John to ride with him in his "fiery chariot . . . away from this sin-sick world" (*Mules and Men,* 70). John, however, escapes from Ole Massa and presumably from his enslavement as well. The message is clear: John, by his superior cunning and speed, evades the presumably-white-sheet-wearing master's ridicule, religion, and ownership. Even the "Kah, Kah, Kah!" (*Mules and Men,* 73) of the audience's laughing response could be read as signifying on the contemporary terrorism of the Ku Klux Klan.

7 Amy Fass Emery, "The Zombie in/as the Text: Zora Neale Hurston's *Tell My Horse,*" *African American Review* 39, no. 3 (2005): 327; Walker, foreword, xii; Hemenway, *Zora Neale Hurston,* 162. Beulah S. Hemingway agrees, stating that Hurston advances a value system based in African rather than white American ideals. Hemingway, "Through the Prism of Africanity: A Preliminary Investigation of Zora Neale Hurston's *Mules and Men,*" in Glassman and Seidel, *Zora in Florida,* 38–45. See also Mary O'Connor, "Zora Neale Hurston and Talking between Cultures," *Canadian Review of American Studies,* special issue, Part I (1992): 141–61, and Susanna Pavloska, *Modern Primitives: Race and Language in Gertrude Stein, Ernest Hemingway, and Zora Neale Hurston* (New York: Garland, 2000), 75–98.

8 Under this admittedly broad rubric, see Houston Baker Jr., "Workings of the Spirit: Conjure and the Space of Black Women's Creativity," in *Zora Neale Hurston: Critical Perspectives Past and Present,* ed. Henry Louis Gates Jr. and K. A. Appiah, 280–308 (New York: Amistad, 1993); Lee Quinby, *Anti-Apocalypse: Exercises in Genealogical Criticism* (Minneapolis: University of Minnesota Press, 1994), 97–112; Susan Edwards

Meisenhelder, *Hitting a Straight Lick with a Crooked Stick: Race and Gender in the Work of Zora Neale Hurston* (Tuscaloosa: University of Alabama Press, 1999); and Leigh Anne Duck, "'Go There tuh *Know* There': Zora Neale Hurston and the Chronotope of the Folk," *American Literary History* 13, no. 2 (2001): 265–94. For two accounts that argue that Hurston empowers black women in particular, see Cheryl A. Wall, "Zora Neale Hurston: Changing Her Own Words," in Gates and Appiah, *Zora Neale Hurston: Critical Perspectives Past and Present,* 76–97, and Mary Katherine Wainwright, "Subversive Female Folk Tellers in *Mules and Men,*" in Glassman and Seidel, *Zora in Florida,* 62–75. For readings that focus on Hurston's investment in the African American oral tradition as an expression of black culture beyond the reach of white appropriation, see Karla F. C. Holloway, *The Character of the Word: The Texts of Zora Neale Hurston* (New York: Greenwood Press, 1987), and Eric J. Sundquist, *Hammers of Creation: Folk Culture in Modern African-American Fiction* (Athens: University of Georgia Press, 1992).

9 Barbara Johnson, "Thresholds of Difference: Structures of Address in Zora Neale Hurston," in *"Race," Writing, and Difference,* ed. Henry Louis Gates Jr. (Chicago: University of Chicago Press, 1986), 323. Other critics who remark on Hurston's troubling of absolute racial distinctions include Elizabeth Fox-Genovese, "My Statue, My Self: Autobiographical Writings of Afro-American Women," in *Reading Black, Reading Feminist: A Critical Anthology,* ed. Henry Louis Gates Jr., 176–203 (New York: Meridian, 1990); Deborah Plant, *Every Tub Must Sit on Its Bottom: The Philosophy and Politics of Zora Neale Hurston* (Urbana: University of Illinois Press, 1995); Samira Kawash, *Dislocating the Color Line: Identity, Hybridity, and Singularity in African American Narrative* (Palo Alto, Calif.: Stanford University Press, 1995); Lynda Marion Hill, *Social Rituals and the Verbal Art of Zora Neale Hurston* (Washington, D.C.: Howard University Press, 1996); Lynn Domina, "'Protection in My Mouf': Self, Voice, and Community in Zora Neale Hurston's *Dust Tracks on a Road* and *Mules and Men,*" *African American Review* 31, no. 2 (1997): 197–209; Ross Posnock, *Color and Culture: Black Writers and the Making of the Modern Intellectual* (Cambridge, Mass.: Harvard University Press, 1998); Delia Caparoso Konzett, *Ethnic Modernisms: Anzia Yezierska, Zora Neale Hurston, Jean Rhys, and the Aesthetics of Dislocation* (New York: Palgrave, 2002); and Werner Sollors, *Ethnic Modernism* (Cambridge, Mass.: Harvard University Press, 2008), esp. 164–67.

10 According to Henry Louis Gates Jr., for example, "the project of finding a voice, . . . selfhood, and empowerment" structures not only Hurston's "oeuvre as a whole," including *Mules and Men,* but also Hurston's own personal circumstance: Hurston is a "masquerader putting on a disguise for the ball, like a character in her fictions. In this way, Hurston *wrote* herself." "Afterword: Zora Neale Hurston: 'A Negro Way of Saying,'" in *Mules and Men,* 289, 294. Gates thus transposes onto Hurston his argument regarding slave narratives (i.e., that authorship is the black person's, especially the black woman's, primary technology of self-formation in a society adamantly opposed to African American autogenesis).

11 Gates, "Afterword," 293, 289.

12 See Rita Felski, *Uses of Literature* (Oxford: Blackwell, 2008), esp. 51–76 and 105–31. Most relevant to my purposes is Felski's claim that both shock and enchantment, though affectively antithetical (113), assault our "sense of autonomy and self-control" (55).

13 Annette Trefzer, "Possessing the Self: Caribbean Identities in Zora Neale Hurston's *Tell My Horse,*" *African American Review* 34, no. 2 (2000): 306. Conversely, Pamela Glenn Menke claims that "voodoo served as more than a metaphor for Hurston. It served as a . . . powerful alternative to . . . a white-inscribed Eurocentric worldview." "'Black Cat Bone and Snake Wisdom': New Orleans Hoodoo, Haitian Voodoo, and Rereading Hurston's *Their Eyes Were Watching God,*" in *Songs of the New South: Writing Contemporary Louisiana,* ed. Suzanne Disheroon Green and Lisa Abney (Westport, Conn.: Greenwood Press, 2001), 123.

14 Zora Neale Hurston, *Tell My Horse* (1938), in *Zora Neale Hurston: Folklore, Memoirs, and Other Writings* (New York: Literary Classics of the United States, 1995), 496; hereinafter cited parenthetically by page number. When Hurston claims that "the symbolism" of voodoo "is taken too seriously," she suggests that it should be recognized as referencing sex, "the infinite, . . . ultimate truth" beyond which "there is no mystery" (*Tell My Horse,* 376). Even if true, however, and it's certainly open to debate, this interpretation is not the kind of political allegory that many critics endorse.

15 See J. C. Dorsainvil, *Vodou et névrose* (1913; repr., Port-au-Prince: Imprimerie la presse, 1931), and *Une Explication Philologique du Vaudou* (Port-au-Prince, 1924); Cyril Scott, *Doctors, Disease, and Health: A Critical Survey of Therapeutics Ancient and Modern* (London: True Health, 1938); and Alfred Metraux, *Haiti: Black Peasants and Voodoo* (New York: Universe Books, 1960), 88–90. Hurston was critical of Dorsainvil's pathologization of voodoo.

16 For "hoodoo doctor[s]" in Hurston's childhood, see her 1942 autobiography *Dust Tracks on a Road,* in *Zora Neale Hurston: Folklore,* 613; hereinafter cited parenthetically by page number. Valerie Boyd contends that "Hurston was a believer. . . . [Her] Eatonville upbringing, along with her explorations of conjure in the summer of 1927, had convinced her that hoodoo was more than magic. . . . It was a secret—and legitimate—religion." Boyd, *Wrapped in Rainbows: The Life of Zora Neale Hurston* (New York: Scribner, 2003), 175. On the prevalence of voodoo in Hurston's writing, see Menke, "Black Cat Bone and Snake Wisdom"; Ellease Southerland, "The Influences of Voodoo on the Fiction of Zora Neale Hurston," in *Sturdy Black Bridges,* ed. Roseann P. Bell, Bettye J. Parker, and Beverly Guy-Sheftall, 171–83 (Garden City, N.Y.: Anchor Books, 1979); Barbara Speisman, "Voodoo as Symbol in *Jonah's Gourd Vine,*" in Glassman and Seidel, *Zora in Florida,* 86–93; and Keith L. Huneycutt, "'The Profound Silence of the Initiated': Zora Neale Hurston's *Polk County,* Dorothy Waring, and Stage Voodoo," in *Proceedings of the 2009 Annual Meeting of the Florida College English Association,* ed. Claudia Slate and Carole Policy, 41–49 (Newcastle upon Tyne, U.K.: Cambridge Scholars, 2010).

17 Peter Kerry Powers, "Gods of Physical Violence, Stopping at Nothing: Masculinity, Religion, and Art in the Work of Zora Neale Hurston," *Religion and American Culture* 12, no. 2 (2002): 231.

18 Though not as popular as *Mules and Men,* contemporary commentators still lauded *Tell My Horse* for giving "the inside story of voodoo." C. G. Woodson, review in Cronin, *Critical Essays,* 145.

19 Somewhat confusingly, Hurston calls New Orleans voodoo "hoodoo" and Haitian Vodou "voodoo." I follow her insistence on the many continuities between them.

20 Trefzer, "Possessing the Self," 310. In this respect, Hurston can be regarded as participating in what Brent Hayes Edwards calls the Harlem Renaissance's "black internationalism." Edwards, *The Practice of Diaspora: Literature, Translation, and the Rise of Black Internationalism* (Cambridge, Mass.: Harvard University Press, 2003).

21 On the "primacy of assemblage" in Vodun (Dahomey) and Vodou (Haiti), see Suzanne Preston Blier, "Vodun: West African Roots of Vodou," in *The Sacred Arts of Haitian Vodou,* ed. Donald J. Consetino (Los Angeles, Calif.: UCLA Fowler Museum of Cultural History, 1995), 75; and Consentino, "Introduction: Imagine Heaven," in Consentino, *Sacred Arts,* 25–55, esp. 27.

22 See Hurston, "Hoodoo in America," *Journal of American Folk-Lore* 44, no. 174 (1931): 317–417.

23 Hurston was critical of uninformed reporting on hoodoo and voodoo throughout her career. See, in particular, her scathing review of "*Voodoo in New Orleans.* By Robert Tallant," *Journal of American Folklore* 60, no. 238 (1947): 436–38. The one work on the subject that she consistently praised was Melville J. Herskovits's *Life in a Haitian Valley* (New York: Alfred A. Knopf, 1937).

24 For studies that have supported Hurston's claims for the African origins of her materials, see Hemingway, "Through the Prism of Africanity," and Keith Cartwright, "'To Walk with the Storm': Oya as the Transformative 'I' of Zora Neale Hurston's Afro-Atlantic Callings," *American Literature* 78, no. 4 (2006): 741–67.

25 Arnold Rampersad, foreword to *Mules and Men,* xx. Hurston regarded Mason as "just as pagan as I" and claimed that "a psychic bond" existed between them (*Dust Tracks,* 688), often referring to Mason as her "angel" in her correspondence. See, for instance, a June 8, 1930, letter to Boas, included in the Franz Boas Papers, Library of Congress, Washington, D.C. Hurston even claimed to have had in childhood a "prophetic vision" of Mason. Zora Neale Hurston, "The Inside Light—Being a Salute to Friendship," in *Zora Neale Hurston: Folklore,* 797. Hurston increasingly felt restrained by Mason's demands, however, and their contractual arrangement was largely over by the time of the publication of *Mules and Men.* For Langston Hughes's criticism of Mason's desire for "black primitivism" from those she patronized (including himself), see *The Big Sea: An Autobiography* (New York: Hill and Wang, 1993), 315.

26 Boyd, *Wrapped in Rainbows,* 157. See also Deborah Plant, *Zora Neale Hurston: A Biography of the Spirit* (Westport, Conn.: Praeger, 2007), 41.

27 In *The Burden of Black Religion* (New York: Oxford University Press, 2008), Curtis J. Evans notes "the irony of the turn to the primitive untouched by the problems of modern society at the precise historical juncture when many African American leaders were seeking to move beyond what they regarded as the backwardness and regressive religion of rural black Southerners" (14).

28 As Joan Dayan asks, "What better way to justify the 'civilizing' presence of marines in Haiti than to project the phantasm of barbarism?" Dayan, "Vodoun, or the Voice of the Gods," *Raritan* 10, no. 3 (1991): 33. See also Dayan, *Haiti, History, and the Gods* (Berkeley: University of California Press, 1995); Wade Davis, *The Serpent and the Rainbow: A Harvard Scientist's Astonishing Journey into the Secret Societies of*

Haitian Voodoo, Zombis, and Magic (New York: Touchstone, 1997); J. Michael Dash, *Haiti and the United States: National Stereotypes and the Literary Imagination* (New York: St. Martin's Press, 1988); and Hans Schmidt, *The United States Occupation of Haiti, 1915–1934* (New Brunswick, N.J.: Rutgers University Press, 1995).

29 For a survey of these popular representations of voodoo, see Joseph M. Murphy, *Working the Spirit: Ceremonies of the African Diaspora* (Boston: Beacon Press, 1994).

30 David Butler's *Jazz Noir: Listening to Music from* Phantom Lady *to* The Last Seduction (Westport, Conn.: Praeger, 2002), 38, "attribute[s] the fascination of the white middle class with voodoo to the 'get rich quick' sensibility and affluence of the Jazz Age, suggesting that the emotional honesty of black music offered an antidote to the 'spiritual vacuity' of post-war America and its growing emphasis on materialism.... White society was merely seeking in jazz a validation of its own intentions, incorrectly perceiving in black culture the hedonistic freedom whites sought for themselves." Butler quotes from Ben Sidran's *Black Talk: How the Music of Black America Created a Radical Alternative to the Values of the Western Literary Tradition* (Edinburgh: Payback Press, 1995), 55.

31 H. I. Brock, "The Full, True Flavor of Life in a Negro Community [Review of *Mules and Men*]" (1935), in Cronin, *Critical Essays,* 40; Harold Courlander, "Witchcraft in the Caribbean Islands [Review of *Tell My Horse*]" (1938), in Cronin, *Critical Essays,* 141. See also Carl Carmer, "In Haiti and Jamaica [Review of *Tell My Horse*]" (1938), in Cronin, *Critical Essays*: "Her sympathies are so strong that she seems to identify herself with her subject" (143).

32 Sheila Walker, translating into explicitly clinical language Katherine Dunham's earlier, implicit suggestion, calls possession a productive "psychotherapeutic technique," even a necessary release of "needs, impulses, and desires... for the sake of the person involved." Walker, *Ceremonial Spirit Possession in Africa and Afro-America: Forms, Meanings, and Functional Significance for Individuals and Social Groups* (Leiden, Netherlands: E. J. Brill, 1972), 86, 88.

33 For the twentieth-century rise of "mail-order companies and hoodoo shops" offering mass-produced versions of such services, see Jeffrey E. Anderson, *Conjure in African American Society* (Baton Rouge: Louisiana State University Press, 2005), 111, and Carolyn Morrow Long, *Spiritual Merchants: Religion, Magic, and Commerce* (Knoxville: University of Tennessee Press, 2001).

34 In one telling example, the death of a man is arranged through his name being written nine times on a piece of paper; the paper being inserted into a dissected cow heart that is then submerged in a jar of vinegar; the jar being placed on another piece of paper with the condemned man's name written on it nine times; and then both jar and paper being placed on an altar that includes a "carved figure of Death" and a toy "coffin" holding "a small man . . . dressed in black to represent the man" (*Mules and Men,* 210). Next, Hurston and her guide dig a grave into which they place a black cat and a black chicken "for a full month" (*Mules and Men,* 210), after which the toy coffin is interred with "their remains" (*Mules and Men,* 211). Then, a "beef brain was placed on a plate with nine hot peppers around it to cause insanity and brain hemorrhages" (*Mules and Men,* 211). Finally, black candles are lit for ninety days, which leads the guide to exclaim, "He cannot live. No one can stand that." It seems he is correct; Hurston notes, simply, "And the man died" (*Mules and Men,* 211). Many of the elements described here are pervasive in hoodoo practice. For the specific use of black candles and dolls for death, see AFS 14268a in the Harry M. Hyatt Collection, Library of Congress, Washington, D.C.

35 For an analogue in contemporary critical thought, see Jane Bennett's "impossibl[e]" attempt "to name the moment of independence . . . possessed by things, a moment that must be there, since things do in fact affect other bodies, enhancing or weakening their power." Bennett, *Vibrant Matter: A Political Ecology of Things* (Durham, N.C.: Duke University Press, 2010), 3.

36 Jean-Luc Nancy explains that "the etymology of 'possessing' . . . reside[s] in the signification of 'being seated on,'" an account that resonates with the idea in Vodou of being mounted by *loa.* Nancy, *Corpus,* trans. Richard A. Rand (New York: Fordham University Press, 2008), 155.

37 Zora Neale Hurston, "How It Feels to Be Colored Me" (1928), in *Zora Neale Hurston: Folklore,* 829. Hurston opens her autobiography with a similar idea: "Like the dead-seeming, cold rocks, I have memories within me that came out of the material that went to make me" (*Dust Tracks,* 561).

38 See Bruno Latour, *Reassembling the Social: An Introduction to Actor-Network-Theory* (Oxford: Oxford University Press, 2007): "to possess is also being possessed; to be attached is to hold and be held" (217).

39 See Maya Deren's claim that in Vodou, "one cannot be man and god at once." Quoted in Murphy, *Working the Spirit,* 41. Hurston records another hoodoo practitioner, "The Frizzly Rooster" (*Mules and Men,*

213), who also "predicted the hour and the minute . . . when he should die" (*Mules and Men,* 214).

40 Leveau's self-displacement redounds even to the recounting of it, as indicated by Hurston's description of Luke Turner, who claimed to be Leveau's nephew and spiritual inheritor: "he wasn't too conscious of me. In fact, he gave me the feeling that he was just speaking, but not for my benefit. He was away off somewhere" (*Mules and Men,* 196). Later, "he sat in a dazed silence" (*Mules and Men,* 198), finally falling into a "hypnotic state" (*Mules and Men,* 204).

41 Rituals involving black cat bones were common in conjure practice, though the desired outcomes differed. One recording at the Library of Congress includes a man explaining that such a bone would grant invisibility to the person who procured it and then placed it in his or her mouth (AFS 1309B LWO 4872 reel 96A). See also "The Trick Bone of a Black Cat," *Journal of American Folk-lore* 12, no. 46 (1899): 228–29.

42 Maya Deren's *The Divine Horseman: The Living Gods of Haiti* (New York: Thames and Hudson, 1953) similarly shuts down while attempting to narrate the possession experience: "The bright darkness floods up through my body, reaches my head, engulfs me. I am sucked down and exploded upward at once. That is all" (260). Against this "strange, subtle thinning out of consciousness" (253), she fights to recover "the certainty of self" and her "proper identity" (256).

43 Boyd, *Wrapped in Rainbows,* 178. Menke notes that cat bones are central to *Jonah's Gourd Vine* (1934) and *Their Eyes Were Watching God* (1937), both written during Hurston's voodoo research. Menke, "Black Cat Bone and Snake Wisdom," 124–25.

44 See Wendy Dutton, "The Problem of Invisibility: Voodoo and Zora Neale Hurston," *Frontiers: A Journal of Woman Studies* 13, no. 2 (1993): 131–52, and Menke, "Black Cat Bone and Snake Wisdom," 127: "This lack of telling confirms Hurston's allegiance to hoodoo/voodoo and her role as voudienne."

45 During an after-hours party at a sawmill, Hurston is confronted by Lucy, who is so jealous of the men's attention toward Hurston that she wants to stab her. The threat triggers a visceral response in Hurston: "I shivered at the thought of dying with a knife in my back, or having my face mutilated" (*Mules and Men,* 154). When the attack finally comes, Hurston exclaims, "I didn't move but I was running in my skin. I could hear the blade already crying in my flesh." She only escapes through the chaos created by the situation: "Curses, oaths,

cries, and the whole place was in motion. Blood was on the floor. I fell out of the door over a man lying on the steps, who either fell himself trying to run or got knocked down. I don't know. I was in the car in a second and in high just too quick" (*Mules and Men,* 179). What is noteworthy about this episode, and the language Hurston uses to record it, is its marked difference from the rest of the book. Unlike the languid prose and settings that characterize most of the first half of *Mules and Men,* this scene dramatizes an immediate danger to Hurston, with her reporting consequently degenerating into impressionistic confusion. What frustrates the ethnographic project, therefore, is not conscious volition but the prospect of physical violation. Interestingly, Hurston links this scene with the Black Cat Bone ritual by having them appear within a page of each other in *Dust Tracks* (699–700).

46 The importance of the materiality of specific textures, times, colors, temperatures, locations, and more is everywhere apparent in the voodoo rituals described by Hurston and others. One representative example comes from a hoodoo man named the Frizzly Rooster: "when you want a person punished who is already indicted, write his name on a slip of paper and put it in a sugar bowl. . . . Now put in some red pepper . . . don't be skeered to put it in, it needs a lot. Put in one eightpenny nail, fifteen cents worth of ammonia and two door keys. You drop one key down in the bowl and you leave the other one against the side of the bowl. . . . Go to your bowl everyday at twelve o'clock and turn the key that is standing against the side of the bowl. That is to keep the man locked in jail. And every time you turn the key, add a little vinegar" (*Mules and Men,* 218). See also recording AFS 1426a in the Harry M. Hyatt Collection, Library of Congress, Washington, D.C.

47 Paul Outka, *Race and Nature from Transcendentalism to the Harlem Renaissance* (New York: Palgrave Macmillan, 2008), 191; Sharon Davie, "Free Mules, Talking Buzzards, and Cracked Plates: The Politics of Dislocation in *Their Eyes Were Watching God,*" *PMLA* 108 (1993): 447.

48 Recall the subtitle of Dorst's essay: "Toward the Death of the Ethnographer."

49 Hemenway, *Zora Neale Hurston,* 248. See also Emery's, "The Zombie in/as the Text," 330, claim that "Hurston's desire to distance herself from voodoo suggests her sense of vulnerability with respect to its powers" and Boyd's, *Wrapped in Rainbows,* 300, assertion that "Hurston decided to retreat from her vigorous investigation of voodoo" while "lying on her sickbed."

50 William James, *Varieties of Religious Experience: A Study in Human Nature: Centenary Edition* (New York: Routledge, 2002), 73. See, though, James's discussion of "the sick soul" and "the lunatics' visions of horror," which are "all drawn from the material of daily fact" (130).

51 James Bissett Pratt, *The Religious Consciousness: A Psychological Study* (1920; repr., New York: Macmillan, 1934), 418, 406, 420, 421, 396.

52 W. T. Stace, *Mysticism and Philosophy* (London: Macmillan, 1960), 161.

53 Rudolf Otto, *The Idea of the Holy: An Inquiry into the Non-rational Factor in the Idea of the Divine and Its Relation to the Rational,* trans. John W. Harvey (1923; repr., Oxford: Oxford University Press, 1958), 13, 12, 13, 25, 13.

54 Ibid., xxi, 62, 13, 62.

55 See Mircea Eliade, *The Sacred and the Profane: The Nature of Religion* (1957; repr., New York: Harcourt, 1987).

56 Nelson Pike, *Mystic Union: An Essay in the Phenomenology of Mysticism* (Ithaca, N.Y.: Cornell University Press, 1992), 27, 165.

57 Cartwright, "To Walk with the Storm," 749, 747, 743, 745; Rampersad, foreword, xxiii. For other accounts of the self- or race-affirming role of voodoo in Hurston's writing, see John Lowe, "Seeing beyond Seeing: Zora Neale Hurston's Religion(s)," *Southern Quarterly* 36, no. 3 (1998): 77–87; Menke, "Black Cat Bone and Snake Wisdom"; Trefzer, "Possessing the Self"; Daphne Lamothe, "Vodou Imagery, African American Tradition, and Cultural Transformation in Zora Neale Hurston's *Their Eyes Were Watching God,*" in *Their Eyes Were Watching God: A Casebook,* ed. Cheryl A. Wall, 165–87 (New York: Oxford University Press, 2000); and Baker, "Workings of the Spirit."

58 Henry Mitchell, *Black Belief: Folk Beliefs in America and West Africa* (New York: Harper and Row, 1975), 53; Murphy, *Working the Spirit,* 40; Walker, *Ceremonial Spirit Possession,* 84. See also Will Coleman, *Tribal Talk: Black Theology, Hermeneutics, and African/American Ways of "Telling the Story"* (University Park: Pennsylvania State University Press, 2000), 14; Joseph R. Washington Jr., *Black Sects and Black Cults* (Garden City, N.Y.: Anchor Books, 1973), 1; and Sterling Stuckey, *Slave Culture: Nationalist Theory and the Foundations of Black America* (New York: Oxford University Press, 1987), 56.

59 Such combinations of affect can occur within a single ceremony: "Instantly the triumphant feeling left the place and was succeeded by one of fear . . . of unspeakable evil. A menace that could not be recognized by ordinary human fears, and the remarkable thing was that everybody seemed to feel it simultaneously and recoiled . . . like a wheat field before wind" (*Tell My Horse,* 411).

60 Departing from more "self-centered" accounts, Deren, *Divine Horsemen,* claims that "possession is not a period of 'self-expression.' On the contrary, the individual's psyche is displaced by that of the loa . . . whose character is constant and independent of that of the person in whose body he becomes manifest" (249). See also Leonard E. Barrett, *Soul-Force: African Heritage in Afro-American Religion* (Garden City, N.Y.: Anchor Press/Doubleday, 1974), 84; Ken Gelder, "Postcolonial Voodoo," *Postcolonial Studies* 3, no. 1 (2000): 89–98; and Russ Castronovo, "The Antislavery Unconscious: Mesmerism, Vodun, and 'Equality,'" *Mississippi Quarterly* 53, no. 1 (1999): 41–56.

61 Edward M. Pavlić makes a similar point when he argues that, in African diasporic religions, the "innovative mixture of de-personalized and inter-personalized processes" results in "internal and social identities [being] disrupted, broken, re-imagined and possibly renewed." Pavlić, "'Papa Legba, Ouvrier Barriere Por Moi Passer': Esu in *Their Eyes* and Zora Neale Hurston's Diasporic Modernism," *African American Review* 38, no. 1 (2004): 67, 66.

62 On the debate over whether Hurston actually supported the occupation, merely signified on it, or, most likely, fell somewhere in between, see, among others, Kevin Meehan, "Decolonizing Ethnography: Zora Neale Hurston in the Caribbean," in *Women at Sea: Travel Writing and the Margins of Caribbean Discourse,* ed. Lizabeth Paravisini-Gebert and Ivette Romero-Cesareo, 245–79 (New York: Palgrave, 2001), esp. 246–48; Trefzer, "Possessing the Self"; and Leigh Anne Duck, "'Rebirth of a Nation': Hurston in Haiti," *Journal of American Folklore* 117, no. 464 (2004): 127–46. Although Hurston's position on the issue in *Tell My Horse* is opaque, in material left out of the original publication of *Dust Tracks,* she forcefully castigates the West for its hypocrisy in decrying Germany's and Japan's invasions while perpetuating their own elsewhere: "the British . . . in India," "our Marines in China," and so on (*Dust Tracks,* 791).

63 See Dayan, *Haiti, History, and the Gods,* esp. 72.

64 Alain Locke argues that Hurston's "gift for poetic phrase, for rare dialect, and folk humor keep her flashing on the surface . . . and from diving down deep either to the inner psychology of characterization or to sharp analysis of the social background." Review of *Their Eyes Were Watching God, Opportunity* 1 (June 1938). Even more harshly, Richard Wright contends that *Their Eyes Were Watching God* trades in the "facile sensuality that has dogged Negro expression since the days of Phillis Wheatley." He continues, "The sensory sweep of her novel carries no theme, no message, no thought. In the main, her novel is

not addressed to the Negro, but to a white audience whose chauvinistic tastes she knows how to satisfy." Wright, "Between Laughter and Tears," *New Masses,* October 5, 1937, 23. See also Harold Preece, "The Negro Folk Cult," *Crisis* 43, no. 12 (1936), 363–74. On the contexts of these attacks, see Hemenway, *Zora Neale Hurston,* and Duck, "Go There tuh *Know* There."

65 Cartwright is thus correct to assert that the "Afro-creole" "spirits" that Hurston conjures "disrupt universalizing conservatisms, liberalisms, and radical 'isms' of Euro-creole hegemony," but he does not see that the autonomous self supposedly thereby liberated is also destabilized in turn. Cartwright, "To Walk with the Storm," 755.

66 Eliade, *Sacred and the Profane,* 13, 12. See also David Brown, *God and Enchantment of Place: Reclaiming Human Experience* (Oxford: Oxford University Press, 2004), 2.

67 Pavlić, "Papa Legba," 80. On "the pluralism of African-based folk religions like hoodoo and voodoo," see also Erik D. Curren, "Should Their Eyes Have Been Watching God? Hurston's Use of Religious Experience and Gothic Horror," *African American Review* 29, no. 1 (1995): 24.

68 Hurston gives the story of a Haitian president whose "best friend" (*Tell My Horse,* 365) was a goat "marri[ed]" to his daughter by a "houngan [who] had mysteriously tied them together" such that "the power of each depended upon the other" (*Tell My Horse,* 363). When the goat dies after the daughter "br[eaks] her vows" (*Tell My Horse,* 365), the president is grief stricken: "It was plain that he considered the goat more than a beast, more than a man, more than just a friend. There was something of worship there" (*Tell My Horse,* 366).

69 Hurston is often dismissive of organized religion, writing in *Dust Tracks* that "people need religion because the great masses fear life and its consequences. . . . Feeling a weakness in the face of great forces, men seek an alliance with omnipotence . . . even though [it] . . . is a creature of their own minds. . . . I feel no need for such" (761–62). Curren, "Should Their Eyes Have Been Watching God?," 22, has argued that her guarded appreciation of voodoo is attributable to its realistic "understanding of nature" over what Hurston calls "organized creeds['] . . . collections of words around a wish" (*Dust Tracks,* 762). Speisman, "Voodoo as Symbol," however, contends that Hurston includes voodoo in her rejection of restrictive religions.

70 Evidence of such agency abounds. Hurston gives numerous examples of animals in particular resisting their assigned roles and assumed passivity, including a defiant mule who quickly disabuses Hurston of her notion that they "might be sisters under the skin" (*Tell My Horse,* 293)

and a goat whose "pathetic, frightened bleat[ing]" voices his "desire to have nothing to do with the . . . grand spectacle of his death" (*Tell My Horse,* 444).

71 Zora Neale Hurston, "What White Publishers Won't Print" (1950), in *Zora Neale Hurston: Folklore,* 953. See also Victor Anderson, *Beyond Ontological Blackness: An Essay on African American Religious and Cultural Criticism* (New York: Continuum, 1995).

72 Maurice Merleau-Ponty, *The Visible and the Invisible,* ed. Claude Lefort, trans. Alphonso Lingis (Evanston, Ill.: Northwestern University Press, 1968), 136.

73 Ibid., 139, 138.

74 Hurston prided herself on taking the first photograph of a zombie and believed that she had solved the mystery of their existence: they were cases of "suspended animation" brought about by "drugs from Africa, which [the] so-called civilized world does not know about." Interview with Mary Margaret McBride on WEAF radio, January 25, 1943, RWC 6858B, Recorded Sound Reference Center, Library of Congress, Washington, D.C. In this, Hurston anticipated the current explanation of zombification, which is that it occurs through the ingestion of tetrodotoxin extracted from puffer fish. See Davis, *Serpent and the Rainbow.*

75 Jacques Derrida, "Structure, Sign, and Play in the Discourse of the Human Sciences," in *Writing and Difference,* trans. Alan Bass (Chicago: University of Chicago Press, 1978), 294; Gilles Deleuze, "Responses to a Series of Questions," *Collapse* 3 (2007): 39. For Derrida's articulation of a nonreligion that "double[s]" or "'haunt[s]' all (positive or historical) religion" in a manner similar to my own imagining of a *philosophical* voodoo, see Hent de Vries, *Philosophy and the Turn to Religion* (Baltimore: Johns Hopkins University Press), x.

76 David Abram, *Becoming Animal: An Earthly Cosmology* (New York: Pantheon, 2010), 27, 313.

77 Ibid., 3, 9. Abram gives token recognition to the fact that "the natural world . . . is hardly a gentle place" (6), but the vast majority of this book and its predecessor, *The Spell of the Sensuous: Perception and Language in a More-Than-Human World* (New York: Vintage, 1997), is devoted to enumerating the undiluted and often explicitly erotic pleasures of other-contact.

78 Trefzer, "Possessing the Self," 305.

79 Pavlić, "Papa Legba," 71.

80 Dana Rush, "Ephemerality and the 'Unfinished' in Vodun Aesthetics," *African Arts* 43, no. 1 (2010): 60.

81 Ibid., 69.

82 Annette Trefzer, "Between Mimesis and Mimicry: Teaching *Tell My Horse*," in *Approaches to Teaching* Their Eyes Were Watching God *and Other Works,* ed. John Lowe (New York: Modern Language Association, 2009), 153–54.

83 Zora Neale Hurston, "Characteristics of Negro Expression," in *Zora Neale Hurston: Folklore,* 835. On the idea that "improvisational codes are the keys to the metaphysical foundations" of diasporic culture, see Paget Henry, "Africana Phenomenology: Its Philosophical Implications," *Worlds and Knowledges Otherwise* 1, no. 3 (2006): 21; Joseph Roach, *Cities of the Dead: Circum-Atlantic Performance* (New York: Columbia University Press, 1996); and Houston Baker Jr., *Blues, Ideology, and Afro-American Literature: A Vernacular Theory* (Chicago: University of Chicago Press, 1984).

84 Pavlić, "Papa Legba," 67, 84.

85 Gerardus van der Leeuw, *Religion in Essence and Manifestation: A Study in Phenomenology,* trans. J. E. Turner (1933; repr., London: George Allen and Unwin, 1938), 465.

86 Martin Heidegger, *Being and Time,* trans. Joan Stambaugh (Albany: State University of New York Press, 1996), 118, 114, 115, 118, 119. Heidegger fails to recognize that *Dasein* is an example of the very "averageness" that he laments as a vitiating "*leveling down* of all possibilities of being" (119). In other words, by limiting the capacity for care to human beings, he, too, "noiselessly squashe[s]" "every exception which thrusts itself to the fore" (119).

87 Ibid., 121, 53, 121. See also van der Leeuw's discussion of the "primitive," to-be-overcome feeling of being "intermingled" with the "environment." Van der Leeuw, *Religion in Essence and Manifestation,* 75.

88 Such a formulation avoids the tendency in "recent postmodern theory" to fetishize "strangerhood as an ontological condition" in which "the stranger . . . *ha[s] a life of its own.*" Sara Ahmed, *Strange Encounters: Embodied Others in Post-coloniality* (New York: Routledge, 2000), 4, 5.

89 In a similar vein, Hurston says regarding an earlier ceremony that it made her "dream strange exalted dreams" that "seemed real for weeks" (*Dust Tracks,* 699).

NOTES TO CODA

1 Elizabeth Grosz rightly notes that such an estranging perspective comes to the fore in modernity with the Darwinian idea that the nonhuman "thing is the precondition of the living and the human." Grosz, "The

Thing," in *The Object Reader,* ed. Fiona Candlin and Raiford Guins (New York: Routledge, 2009), 125.

2 I do not include Douglass and Equiano in this summary because, as I argue in chapter 3, their overcoming of the metaphysics of slavery to become men is—by their own proclamation—the essential story of modernity. Of course, their *achievement* of humanity illuminates modernity's hypocrisy and terrible cost (even unto the world itself), but it would be disingenuous to present their work—rather than the diasporic cosmologies that they reject—as posthumanist.

3 I borrow this language from Timothy Morton's discussion of "ecology without holism" in *The Ecological Thought* (Cambridge, Mass.: Harvard University Press, 2010), 127.

4 Alfred North Whitehead, *Modes of Thought* (1938; repr., New York: Free Press, 1966), 174.

5 Cary Wolfe, *Critical Environments: Postmodern Theory and the Pragmatics of the "Outside"* (Minneapolis: University of Minnesota Press, 1998), 41.

6 On "texts as 'nonhuman actors,'" see Rita Felski, "Context Stinks!" *New Literary History* 42, no. 4 (2011): 574.

7 Bruno Latour, *The Pasteurization of France,* trans. Alan Sheridan and John Law (Cambridge, Mass.: Harvard University Press, 1988), 220.

8 For Castiglia's discussion of a "humanism without humans," by which he means a politics that "privileges rational deliberation" over appeals to the individual subject's "feelings and desire[s]," see *Interior States: Institutional Consciousness and the Inner Life of Democracy in the Antebellum United States* (Durham, N.C.: Duke University Press, 2008), 294, 297. My "posthumanism without posthumans" follows this inquiry into a politics that is not based in "the 'self' . . . as a foundational and autonomous ground" but that departs from Castiglia's faith in the "deliberative function" as a comprehensive alternative (296).

9 Bruno Latour, *Pandora's Hope: Essays on the Reality of Science Studies* (Cambridge, Mass.: Harvard University Press, 1999), 281.

10 Gayatri Chakravorty Spivak, *Death of a Discipline* (New York: Columbia University Press, 2003), 71; Ursula K. Heise, *Sense of Place and Sense of Planet: The Environmental Imagination of the Global* (New York: Oxford University Press, 2008), 56; William E. Connolly, *A World of Becoming* (Durham, N.C.: Duke University Press, 2011). See also Wai Chee Dimock's attempt to "think of the planet as a plausible whole" in *Through Other Continents: American Literature across Deep Time* (Princeton, N.J.: Princeton University Press, 2006), 5.

11 Zora Neale Hurston, "Seeing the World As It Is," in *Zora Neale Hurston: Folklore, Memoirs, and Other Writings* (New York: Literary Classics of the United States, 1995), 787.

12 See Thomas L. Dumm, "The Problem of 'We'; or, The Persistence of Sovereignty," *boundary 2* 26, no. 3 (1999): 55–61.

13 Hurston, "Seeing the World As It Is," 786.

14 Levi R. Bryant, *The Democracy of Objects* (Ann Arbor, Mich.: Open Humanities Press, 2011); Bruno Latour, "From Realpolitik to *Dingpolitik* or How to Make Things Public," in Candlin and Guins, *Object Reader,* 154; Latour, *We Have Never Been Modern,* trans. Catherine Porter (Cambridge, Mass.: Harvard University Press, 1993), 144; Jane Bennett, *Vibrant Matter: A Political Ecology of Things* (Durham, N.C.: Duke University Press, 2010), xix; Isabelle Stengers, *Cosmopolitics I,* trans. Robert Bononno (Minneapolis: University of Minnesota Press, 2010), and Stengers, *Cosmopolitics II,* trans. Robert Bononno (Minneapolis: University of Minnesota Press, 2011). We might also add to this list Morton's attempt to imagine "complex kinds of democracy that take the nonhuman into account" in *Ecological Thought,* 126; the collection *Political Matter: Technoscience, Democracy, and Public Life,* ed. Bruce Braun and Sarah J. Whatmore (Minneapolis: University of Minnesota Press, 2010); and Christopher Toulmin's *Cosmopolis: The Hidden Agenda of Modernity* (Chicago: University of Chicago Press, 1992).

15 Isabelle Stengers, "Including Nonhumans in Political Theory: Opening Pandora's Box?," in Braun and Whatmore, *Political Matter,* 3–33. Another version of this "(re)invention of politics" is advanced in "The Cosmopolitical Question," chapter 23 of Stengers, *Cosmopolitics II,* 352.

16 Latour, *We Have Never Been Modern,* 29; Latour, *Reassembling the Social: An Introduction to Actor-Network-Theory* (Oxford: Oxford University Press, 2007), 254.

17 Bruno Latour, *Politics of Nature: How to Bring the Sciences into Democracy,* trans. Catherine Porter (Cambridge, Mass.: Harvard University Press, 2004), 231; Latour, *Reassembling the Social,* 254.

18 Latour thus amplifies the stakes of Spivak's earlier posing of the question of "How many are we?," which refers only to humans. Spivak, *Death of a Discipline,* 102.

19 Latour, *Politics of Nature,* 232.

20 Bruno Latour, "On Recalling ANT," in *Actor Network Theory and After,* ed. John Law and John Hassard (Oxford: Blackwell, 2007), 20. This insistence on the primacy of relationality over the things relating is the

site at which Latour departs most dramatically from object-oriented ontology, which holds that "metaphysics demands a distinction between the inner reality of a thing and its relations with other entities." Graham Harman, *Prince of Networks: Bruno Latour and Metaphysics* (Melbourne: re.press, 2009), 75.

21 Colin Barron, ed., "A Strong Distinction between Humans and Nonhumans Is No Longer Needed for Research Purposes: A Debate between Bruno Latour and Steve Fuller," *History of the Human Sciences* 16, no. 2 (2003): 92.

22 Latour, *Politics of Nature,* 39.

23 Barron, "A Strong Distinction," 79.

24 Latour, *Reassembling the Social,* 259.

25 Ibid.

26 Latour, *We Have Never Been Modern,* 145.

27 Latour, *Reassembling the Social,* 235.

28 Latour, *We Have Never Been Modern,* 136.

29 As Cary Wolfe notes, "one of the hallmarks of humanism . . . is its penchant for that kind of pluralism . . . in which the sphere of attention and consideration (intellectual or ethical) is broadened and extended to the previously marginalized groups, but without in the least destabilizing . . . the schema of the human who undertakes such pluralism. In that event, pluralism becomes incorporation, and the projects of humanism (intellectually) and liberalism (politically) are extended, and indeed extended in a rather classic sort of way." Wolfe, *What Is Posthumanism?* (Minneapolis: University of Minnesota Press, 2010), 99.

30 See Mary Gabriel, *Notorious Victoria: The Life of Victoria Woodhull, Uncensored* (Chapel Hill, N.C.: Algonquin Books, 1998), which includes a copy of the "cosmopolitical party platform" (303–7), and Barbara Goldsmith, *Other Powers: The Age of Suffrage, Spiritualism, and the Scandalous Victoria Woodhull* (New York: HarperPerennial, 1999). Woodhull's Cosmopolitical Party could be seen as the conceptual opposite of the "Cosmos Club," a social and intellectual group cofounded by Henry Adams in 1878 for the white male elites of Washington, D.C.

31 Simon Schaffer, "The Eighteenth Brumaire of Bruno Latour," *Studies in History and Philosophy of Science* 22, no. 1 (1991): 182.

32 David Bloor, "Anti-Latour," *Studies in the History and Philosophy of Science* 30, no. 1 (1999): 82; Barron, "A Strong Distinction," 82. For an extended version of Steve Fuller's criticism of Latour's approach as being "user-friendly" to capitalism because of its lack of meaningful commitments, see "Why Science Studies Has Never Been Critical of Science: Some Recent Lessons on How to Be a Helpful Nuisance and a

Harmless Radical," *Philosophy of the Social Sciences* 30, no. 1 (2000): 5.

33 Latour, "From Realpolitik to *Ding*politik," 155.

34 Latour, *We Have Never Been Modern,* 136, 138.

35 Latour, "On Recalling ANT," 16.

36 Ibid., 18.

37 See Charles Taylor, K. Anthony Appiah, Jürgen Habermas, Steven C. Rockefeller, Michael Walzer, and Susan Wolf, *Multiculturalism: Examining the Politics of Recognition,* ed. Amy Gutmann (Princeton, N.J.: Princeton University Press, 1994), and Walter Benn Michaels, *The Trouble with Diversity: How We Learned to Love Identity and Ignore Inequality* (New York: Metropolitan Books, 2006). I disagree with Michaels's assertion that the politics of recognition is separable from the politics of redistribution, but I also do not think that the two are interchangeable, as Latour's quasi-metaphysics occasionally appears to suggest.

38 Barron, "A Strong Distinction," 93.

39 Ibid., 93, 81.

40 Stengers, "Including Nonhumans," 3; Latour, *Reassembling the Social,* 261.

41 Latour, *Politics of Nature,* 235. Latour himself directs investigators to "go slow" in their transition from counting to assembling in *Reassembling the Social,* 190; I'm only recommending an even longer dilation. See also Stengers's repeated insistence on the necessity of slowness and hesitation in matters of critical judgment. Stengers, *Cosmopolitics I,* 73, 179, 229; Stengers, "Including Nonhumans," 22.

42 In *Giving an Account of Oneself* (New York: Fordham University Press, 2005), Judith Butler, drawing on Levinas, Adorno, and Foucault, articulates an ethics based in the relationality of mutual vulnerability rather than the isolated, "self-sufficient 'I'" (136), which "cannot fully furnish the grounds of its own emergence" (116). Related arguments are offered by Jacques Derrida, "The Animal That Therefore I Am," in *The Animal That Therefore I Am,* ed. Marie-Louise Mallet, trans. David Wills (New York: Fordham University Press, 2008), 28: the "nonpower at the heart of power"; Wolfe, *What Is Posthumanism?,* 139: a "shared finitude, a shared 'passivity' . . . that runs counter to the liberal model of the subject as above all a creature of volition, autonomy, and agency"; and Morton, *Ecological Thought,* 127: "a collectivity of weakness, vulnerability, and incompletion." See also Thomas Carl Wall, *Radical Passivity: Levinas, Blanchot, and Agamben* (Albany: State University of New York Press, 1999).

43 Barbara Johnson argues that an ethics of refraining from violent appropriation presupposes the very thing it wants to avoid: an inequality

of power between self and other. Johnson, *Persons and Things* (Cambridge, Mass.: Harvard University Press, 2008), 94. However, she fails to acknowledge the recursive *mutuality* of vulnerability, as indicated by her Lacanian recommendation to violently use the world to realize its capacity to survive us, which does not seem like a viable option in an age of mass extinctions.

44 Christopher Hitt, "Toward an Ecological Sublime," *New Literary History* 30, no. 3 (1999): 611.

45 Isabelle Stengers, with Ilya Prigogine, "The Reenchantment of the World," in *Power and Invention: Situating Science,* trans. Paul Bains (Minneapolis: University of Minnesota Press, 1997), 40.

INDEX

MATTHEW A. TAYLOR is an assistant professor in the Department of English and Comparative Literature at the University of North Carolina–Chapel Hill.